National Aeronautics and Space Administration

THE SATURN SYSTEM
THROUGH THE EYES OF CASSINI

TABLE OF CONTENTS

THE SATURN SYSTEM
THROUGH THE EYES OF CASSINI

This book was developed collaboratively by the National Aeronautics and Space Administration (NASA) including NASA's Planetary Science Division (PSD), NASA's Jet Propulsion Laboratory (JPL), and the Lunar and Planetary Institute (LPI), operated for NASA by Universities Space Research Association.

ON THE COVER: **Nested Rings**

Saturn's northern hemisphere is seen here against its nested rings. This view from the Cassini spacecraft looks toward the unilluminated side of the rings from about 30 degrees above the ring plane. The rings have been brightened relative to the planet to enhance visibility. Images taken using red, green, and blue spectral filters were combined to create this natural-color view. Cassini captured these images on February 24, 2009, at approximately 538,000 miles (866,000 kilometers) from Saturn. Image Credit: NASA/JPL-Caltech/Space Science Institute

Published 2017

For complete media usage guidelines, please visit https://www.nasa.gov/multimedia/guidelines

This publication is available as a free download at https://www.nasa.gov/ebooks

FOREWORD
BY JAMES GREEN

DIRECTOR,
NASA
PLANETARY SCIENCE DIVISION

JULY 2017

More than 400 years ago, Galileo Galilei trained his homemade telescope on the night sky and observed that Saturn had two objects closely related to the planet extending on either side. At the time, in 1610, Galileo declared them to be moons.

A few decades later, Saturn moon science accelerated at a dizzying pace. Christiaan Huygens first observed Saturn's largest moon Titan in 1655 and was the first to describe the extended "moon-like" features at Saturn as a disk of material surrounding the planet. From 1671 to 1674, Giovanni Cassini discovered the moons Iapetus, Rhea, Dione and Tethys. In 1675, Cassini discovered the gap in Saturn's rings that we now know as the "Cassini Division."

In the space age, before the Cassini-Huygens mission, we had only hints of the discoveries awaiting us at Saturn. Pioneer 11 and Voyagers 1 and 2 conducted flybys decades ago. But these quick encounters didn't allow time for more extensive research. NASA and the European Space Agency created a partnership to orbit a Saturn orbiter (Cassini) and a lander (Huygens) on Titan. Like its namesakes, the Cassini-Huygens mission not only discovered previously unknown moons, but it also helped us understand the science behind their formation, their interactions with the rings, and how truly diverse they are.

The Cassini-Huygens mission revolutionized what we know about the Saturn system. The rings of Saturn, the moons, and the planet itself offer irresistible and inexhaustible subjects for intense study, and Cassini-Huygens did not disappoint. The Saturnian system proved to be a rich ground for science exploration and discoveries, and Cassini has been nothing short of a discovery machine.

At the time Cassini plunged into Saturn at the end of its mission, it had observed the planet for a little less than half of a Saturn year. Cassini orbited the gas giant 294 times, traveling 4.9 billion miles and collecting 635 gigabytes (GB) of science data, forever changing our knowledge of the Saturn system and yielding tremendous insight for understanding the entire Solar System.

Cassini's observations have given us new views of the planet that provided a plethora of iconic images. The mission has made groundbreaking discoveries in our relentless search for life in the Solar System — whether revealing that Titan is going through similar cycles as Earth before life evolved, or spying icy plumes on Enceladus. In addition, Cassini's 13-year mission made it possible to watch changes in Saturn's dynamic ring system and observe what may be one of the most active, chaotic rings in our solar system — Saturn's F ring. The spacecraft discovered propeller-like formations in the rings, witnessing the possible birth of a new moon.

Since its arrival in 2004, Cassini-Huygens has astounded us with data and images never before obtained with such detail and clarity. This book is the tip of a planet-sized iceberg. Over the last 13 years, Cassini has taken about 450,000 spectacular images within the Saturn system. How can we pare down this number to just a few when each snapshot, each image, tells a new story? The sheer beauty of these pictures is surpassed only by the science and discoveries they represent. Cassini taught us that Saturn is a far cry from a tranquil lone planet with delicate rings. Now, we know more about Saturn's chaotic, active, and powerful rings, and the storms that rage beneath. Some of the Titan and Enceladus images hint at the possibility of life never before suspected. And how could we forget those exciting first looks at the moons we never knew existed?

Each of the images in this book is representative of so many more, and reflects both astounding beauty and the new science that Cassini-Huygens pioneered. Most of these images come from the hard work of the Imaging Science Subsystem Team, which provided special processing. I would like to thank them personally and all the other Cassini team members for a job well done.

So, congratulations Cassini, for forging our path. This book is the first chapter of what I predict will be the greatest story ever told: how humans reached for the stars and discovered life beyond Earth. I am honored to be part of it.

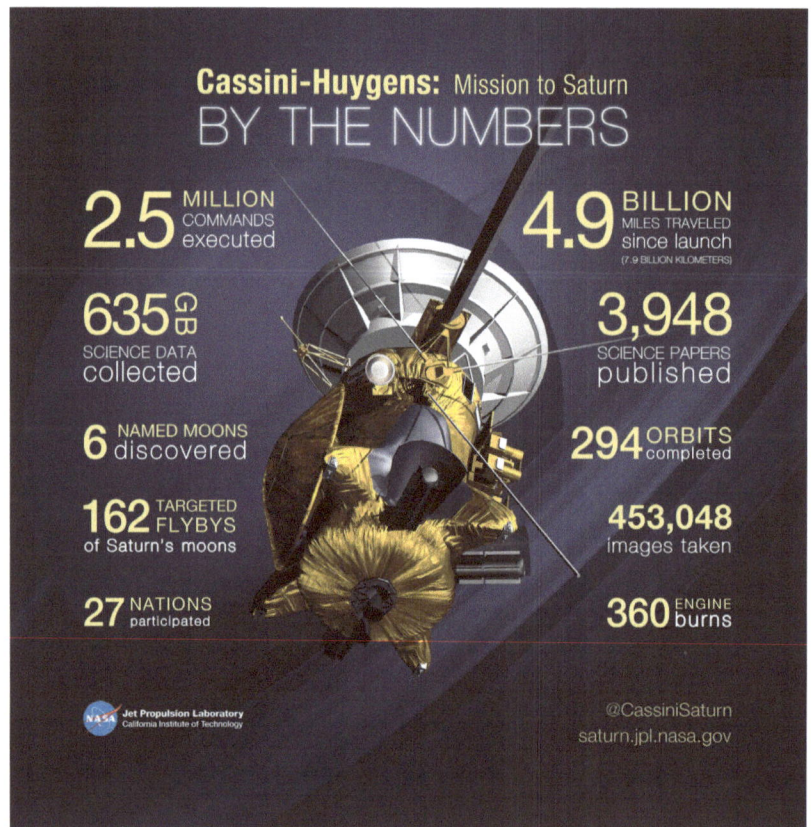

A snapshot of some of the impressive numbers Cassini amassed during its 20-year mission.

1

SATURN:
CROWN JEWEL OF THE
SOLAR SYSTEM

In the 400 years since humans began studying the cosmos with telescopes, Saturn has been transformed from a faint and blurry spot in the sky to a colossal planet in sharp focus. Yet, it seems that the clearer we see Saturn, the more enigmas we find.

Saturn is 763 times the volume of Earth with a primarily hydrogen and helium atmosphere that's like a wild, churning cauldron. "Most of the weather on Saturn is erratic," said Andrew Ingersoll, a member of the Cassini imaging team and an atmospheric physicist at Caltech. "A year will go by with no thunderstorms at all, but every 20 to 30 years you get a giant storm flashing with lightning every tenth of a second." Cassini watched that mega-storm erupt in 2010, sending atmospheric disturbances all the way around the planet until the storm began running into its own wake, appearing to eat its own tail before it faded and disappeared.

Not all of Saturn's characteristics come and go, though. "Some features are surprisingly long-lived," Ingersoll said. "Cassini observed Saturn's north polar hexagon, discovered in 1981 by Voyager, and it's big enough to contain two Earths. And each pole sports a hurricane-like 'eye' that has been around for a decade or more. These are special properties, unique to Saturn, and they are showing us how extreme the weather can be."

Further, scientists don't know with precision how long it takes for Saturn to complete a single rotation on its axis (i.e., the length of a Saturn day), or why the planet's magnetic poles are aligned with the axis of rotation — the only planet known to have such an alignment. The size of Saturn's core and the planet's internal structure are also practically unknown.

Cassini's 13 years in Saturn orbit were the first opportunity to study this turbulent planet up close for years on end, and the spacecraft revealed new insights into Saturn's puzzles. Though a Saturn year lasts nearly 30 Earth years, Cassini orbited for at least part of three Saturnian seasons, during which the planet changed and so did our knowledge of this giant world. As a result of Cassini's years of close study of Saturn, we now have countless images and other data about this surreal planet and its accompanying rings and moons.

These images serve merely as a collective portrait or snapshot of Saturn, but this ever-changing world will always offer more to see.

Chapter Title Image: **Ringworld Waiting**

Saturn's peaceful beauty invites the Cassini spacecraft for a closer look in this natural-color view, taken during the spacecraft's approach to the planet. By this point in the approach sequence, Saturn was large enough that two narrow-angle camera images were required to capture an end-to-end view of the planet, its delicate rings, and several of its icy moons. The composite is made entirely from these two images. Moons visible in this mosaic: Epimetheus (72 miles, 116 kilometers across), Pandora (52 miles, 84 kilometers across), and Mimas (247 miles, 398 kilometers across) at left of Saturn; Prometheus (63 miles, 102 kilometers across), Janus (113 miles, 181 kilometers across), and Enceladus (310 miles, 499 kilometers across) at right of Saturn. Image Credit: NASA/JPL-Caltech/Space Science Institute

WAVES AND SHEAR

This remarkably detailed view of Saturn's clouds reveals waves at the northern boundary of the bright equatorial zone, presumably associated both with the strong wind shear there and also the difference in density across the boundary with the band to the north. To the south, two dark ovals embrace, while dark ring shadows blanket the north. The moon Janus (113 miles, 181 kilometers across) occupies a mere two pixels beneath the rings, at right of center. Cassini captured this image on March 16, 2006, approximately 1.3 million miles (2 million kilometers) from Saturn. Image Credit: NASA/JPL-Caltech/Space Science Institute

NESTED RINGS

Saturn's northern hemisphere is seen here against its nested rings. This view from the Cassini spacecraft looks toward the unilluminated side of the rings from about 30 degrees above the ring plane. The rings have been brightened relative to the planet to enhance visibility. Images taken using red, green, and blue spectral filters were combined to create this natural-color view. Cassini captured these images on February 24, 2009, at approximately 538,000 miles (866,000 kilometers) from Saturn. Image Credit: NASA/JPL-Caltech/Space Science Institute

SUN-STRIPED SATURN

In a dazzling and dramatic portrait painted by the Sun, the long thin shadows of Saturn's rings sweep across the planet's northern latitudes. Within the shadows, bright bands represent areas where the ring material is less dense, while dark strips and wave patterns reveal areas of denser material. The shadow darkens sharply near upper right, corresponding to the boundary of the thin C ring with the denser B ring. The globe of Saturn's moon Mimas (247 miles, or 398 kilometers, across) has wandered into view near the bottom of the frame. A few large craters are visible on the small moon. Cassini captured this image on January 18, 2005, approximately 889,000 miles (1.4 million kilometers) from Saturn. Image Credit: NASA/JPL-Caltech/Space Science Institute

LIMB SCAN

This bizarre scene shows the cloud-streaked limb of Saturn in front of the planet's B ring. The ring's image is warped near the limb by the diffuse gas in Saturn's upper atmosphere. This view looks toward the sunlit side of the rings from about 31 degrees below the ring plane. North on Saturn is up. Cassini captured this image on June 24, 2008, at approximately 408,000 miles (657,000 kilometers) from Saturn using a spectral filter sensitive to wavelengths of infrared light centered at 750 nanometers. Image Credit: NASA/JPL-Caltech/Space Science Institute

THE RITE OF SPRING

Saturn has seen countless equinoxes since the birth of the solar system, but this is the first witnessed up close by Cassini. From 20 degrees above the ring plane, Cassini shot 75 exposures in succession for this mosaic showing Saturn, its rings, and a few of its moons a day and a half after Saturn equinox on August 12, 2009, when the sun's disk was exactly overhead at the planet's equator. These scenes are possible only during the few months before and after Saturn's equinox, which occurs only once approximately every 15 Earth years. Cassini captured these images at approximately 526,000 miles (847,000 kilometers) from Saturn. Image Credit: NASA/ JPL-Caltech/Space Science Institute

BEFORE WIDE SHADOWS

Saturn's moon, Tethys, orbits in front of the wide shadows cast by the rings onto the planet. Tethys (660 miles, or 1,062 kilometers across) appears just below the rings near the center of the image. This view looks toward the northern, sunlit side of the rings from less than one degree above the ring plane. Cassini captured this image on December 7, 2011, at approximately 1.1 million miles (1.8 million kilometers) from Tethys using a spectral filter sensitive to wavelengths of near-infrared light centered at 752 nanometers. Image Credit: NASA/JPL-Caltech/Space Science Institute

HEXAGON AND RINGS

Saturn's north polar hexagon basks in the Sun's light once spring came to the northern hemisphere. Many smaller storms dot the north polar region and Saturn's signature rings, which seem to disappear because of Saturn's shadow. Cassini captured this image on November 27, 2012, using a spectral filter sensitive to wavelengths of near-infrared light centered at 750 nanometers. The view was acquired at a distance of approximately 403,000 miles (649,000 kilometers) from Saturn. Image Credit: NASA/JPL-Caltech/Space Science Institute

JEWEL OF THE SOLAR SYSTEM

Cassini swung high above Saturn to reveal this stately view of the golden-hued planet and its main rings. The view is in natural color, as human eyes would have seen it. This mosaic was made from 36 images in three color filters obtained by Cassini on October 10, 2013. Saturn sports differently colored bands of weather in this image. For instance, a bright, narrow wave of clouds around 42 degrees north latitude appears to be some of the turbulent aftermath of a giant storm that reached its violent peak in early 2011. The mysterious six-sided weather pattern known as the hexagon is visible around Saturn's north pole. Image Credit: NASA/JPL-Caltech/SSI/Cornell

BLINDING
SATURN

Saturn sits enveloped by the full splendor of its stately rings. Taking in the rings in their entirety was the focus of this particular imaging sequence. The camera exposure times were just right to capture the dark-side of its rings, but longer than that required to properly expose the globe of sunlit Saturn. Consequently, the sunlit half of the planet is overexposed. A strip of twilight on the globe displays colorful details in the atmosphere. Bright clouds dot the bluish-grey northern polar region here. In the south, the planet's night side glows golden in reflected light from the rings' sunlit face. Cassini captured this image on January 19, 2007, approximately 764,000 miles (1.23 million kilometers) from Saturn. Image Credit: NASA/JPL-Caltech/Space Science Institute

INFRARED

This false-color composite image made from 65 individual Cassini observations shows Saturn's rings and southern hemisphere on November 1, 2008. In this image constructed from data collected in the near-infrared wavelengths of light, scientists designated blue to indicate sunlight reflected at a wavelength of 2 microns, green to indicate sunlight reflected at 3 microns, and red to indicate thermal emission at 5 microns. The heat emission from the interior of Saturn is only seen at 5 microns wavelength in the spectrometer data, and thus appears red. Image Credit: NASA/JPL-Caltech/ASI/University of Arizona

CATCHING
ITS TAIL

Cassini gives us this true-color view of the largest, most intense storm observed on Saturn. The image was captured on February 25, 2011, when the storm had formed a tail that wrapped around the entire planet. Some of the clouds moved south and got caught up in a current that flows to the east (to the right) relative to the storm head. This tail, which appears as slightly blue clouds south and west (left) of the storm head, can be seen encountering the storm head in this view. The images were taken approximately 1.4 million miles (2.2 million kilometers) from Saturn. Image Credit: NASA/JPL-Caltech/Space Science Institute

'TIS THE SEASON

Winter is approaching in the southern hemisphere of Saturn in this view, captured by Cassini on July 29, 2013, at approximately 1 million miles (1.6 million kilometers) from Saturn. The changing blue hue that we have learned marks winter at Saturn is likely due to reduction of ultraviolet sunlight and the haze it produces, making the atmosphere clearer and increasing the opportunity for Rayleigh scattering (scattering by molecules and smaller particles) and methane absorption — both processes make the atmosphere blue. The small black dot seen to the right and up from image center, within the ring shadows of the A and F rings, is the shadow of the moon, Prometheus. Image Credit: NASA/JPL-Caltech/Space Science Institute

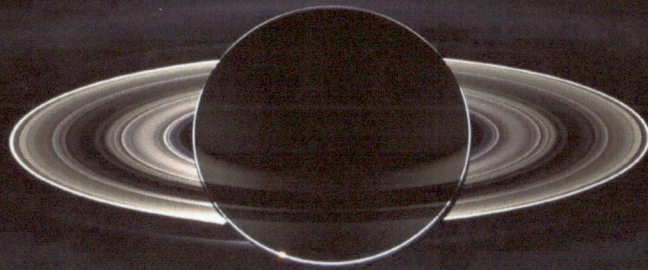

IN SATURN'S SHADOW

With giant Saturn hanging in the blackness and sheltering Cassini from the sun's blinding glare, the spacecraft viewed the rings as never before, revealing previously unknown faint rings and even glimpsing its home world. This marvelous panoramic view was created by combining 165 images taken by Cassini on September 15, 2006, as the spacecraft drifted in the darkness of Saturn's shadow for about 12 hours, allowing a multitude of unique observations of the microscopic particles that compose Saturn's faint rings. Color in the view was created by digitally compositing ultraviolet, infrared, and clear filter images and was then adjusted to resemble natural color. Cassini took the images at approximately 1.3 million miles (2.2 million kilometers) from Saturn. Image Credit: NASA/JPL-Caltech/Space Science Institute

2

RINGS:
ICE PARTICLES, MOONLETS, AND GRAVITY

No other planet in our solar system has rings as splendid or spectacular as Saturn's. So expansive and bright are Saturn's rings that they were discovered as soon as humans began pointing telescopes at the night sky. Galileo Galilei was the first person known to study the heavens through a telescope, and he secured a place in an astronomical history when he discovered Jupiter's four largest moons in 1610. Despite Saturn being roughly twice as far from Earth as Jupiter, Saturn's rings are so big and brilliant that Galileo observed them in the same year he spotted Jupiter's moons.

Galileo didn't exactly understand what he was looking at, and in the centuries since their discovery, Saturn's rings have remained to some extent a puzzle. Scientists have long tried to understand the exact composition of the rings, as well as their age and origin. Scientists don't even know their mass precisely. Cassini has worked to answer those questions.

Despite Saturn being the only planet in our solar system to possess rings of such magnitude, rings have nonetheless become popularly synonymous with the idea of a planet. "Saturn's rings are truly extraordinary," said Cassini Project Scientist Linda Spilker. "They're actually made up of separate ice particles, and the thought that millions of individual particles, each on its own path, can combine to create such intricate, beautiful waves and structures is astonishing."

Spilker says that up close, the ring particles might look like fluffy snowballs, but that inside they might be more like solid ice, and they vary extremely in size. "Some are only the size of tiny marbles while others are the size of mountains," she said. Spilker has no trouble imagining being

a spacesuited human floating along with the ring particles in their orbits. "You would gently collide with the particles, and some might even stick to your spacesuit," Spilker said. "They are in a thin layer too, so you would be able to easily dive through the particles to the other side of the rings."

Because Cassini was the first spacecraft to orbit Saturn, and did so for more than 13 years, Cassini documented the rings as no spacecraft could before, discovering previously unknown features and behaviors, some of which are included in the following pages. Even through the mission's final months, the Cassini family still gasped when viewing the spacecraft's latest images of Saturn's rings.

Chapter Title Image: **Colorful Division**

The rings are awash in subtle tones of gold and cream in this view, which looks toward the unlit side of the rings from about 30 degrees above the ring plane and shows the outer B ring, the Cassini Division, and the inner part of the A ring. This natural-color view was created from images taken using red, green, and blue spectral filters captured by Cassini on September 29, 2006, at approximately 1.14 million miles (1.83 million kilometers) from Saturn. Image Credit: NASA/JPL-Caltech/Space Science Institute

PRETTY IN PINK

Saturn and its rings are prominently shown in this color image, along with three of Saturn's smaller moons, which are (left to right) Prometheus, Pandora, and Janus. Prometheus and Pandora are often called the "F ring shepherds" because they control and interact with Saturn's F ring, seen between them. This image was created by combining images the Cassini spacecraft took using red, green, and blue filters on June 18, 2004, from 5.1 million miles (8.2 million kilometers) from Saturn. Contrast has been enhanced to aid visibility. Image Credit: NASA/JPL-Caltech/ Space Science Institute

SOUTHERN COLOR

The Cassini spacecraft samples a bit of Saturn's southern hemisphere along with a spread of the planet's main rings. Working outward from the planet, the C, B, and A rings are visible in this image. The rings have been brightened relative to the planet to enhance their visibility. This natural-color view looks toward the sunlit side of the rings from about 59 degrees below the ring plane. The image was created by combining images taken with red, green, and blue spectral filters taken by Cassini on April 23, 2009, at a distance of approximately 621,000 miles (1 million kilometers) from Saturn. Image Credit: NASA/JPL-Caltech/Space Science Institute

ACROSS RESPLENDENT RINGS

The shadow of Saturn's moon, Mimas, dips onto the planet's rings and straddles the Cassini Division in this natural-color image taken as Saturn approaches its August 2009 equinox, which occurs only once about every 15 Earth years. The illumination geometry in the few months before and after equinox allows moons orbiting in or near the plane of Saturn's equatorial rings to cast shadows onto the rings. This view looks toward the sunlit side of the rings from about 52 degrees below the ring plane and was created by combining images taken by Cassini using red, green, and blue spectral filters on April 8, 2009, at a distance of approximately 684,000 miles (1.1 million kilometers) from Saturn. Image Credit: NASA/JPL-Caltech/Space Science Institute

RHEA'S RING SHADOW

Rhea joins other Saturnian moons in casting a shadow on the rings in this image taken as Saturn approached its August 2009 equinox. The night side of the planet is dimly illuminated here by ringshine, the southern hemisphere more so than the north. The excess brightness in the lower left of the image is lens flare, an artifact from light scattering within the camera optics. This view looks toward the unilluminated side of the rings from about 17 degrees above the ring plane. Cassini captured this image in visible light on July 21, 2009, from approximately 1.2 million miles (1.9 million kilometers) from Saturn. Image Credit: NASA/JPL-Caltech/Space Science Institute

GLORIOUS VIEW

Sunlight took many different paths to compose this glorious image of Saturn and its rings. Sunlight reflects off the illuminated side of the rings to light the planet's southern hemisphere. The northern hemisphere (top left corner of the image) is dimly lit by light scattered through the rings. The planet's shadow cuts across the rings, but light reflected off the southern hemisphere backlights parts of the C ring, making them visible in silhouette. Stars occulted by the rings make bright points of light in the image. This natural-color view was created by combining images taken by Cassini on March 20, 2009, using red, green, and blue spectral filters at approximately 554,000 miles (892,000 kilometers) from Saturn. Image Credit: NASA/JPL-Caltech/Space Science Institute

CRISS-CROSSED RINGS

Saturn's rings appear to intersect themselves in an impossible way in this Cassini image, but the image actually shows the rings in front of the planet, upon which the shadow of the rings is cast. Because rings like the A ring and Cassini Division, in the foreground, are not entirely opaque, the disk of Saturn and those ring shadows can be seen directly through the rings themselves. Saturn's moon, Pan (17 miles or 28 kilometers across, near image center), keeps open the Encke gap. Cassini took this image in visible light on February 11, 2016, at approximately 1.2 million miles (1.9 million kilometers) from Pan. Image Credit: NASA/JPL-Caltech/Space Science Institute

VEIL OF ICE

The Cassini spacecraft stares toward Saturn through its gauzy veil of rings. The great ice-particle screen acts like a filter, attenuating the glare from the planet and making its high-altitude haze easy to see. The F ring shows off the faint ringlets flanking its core, and a single ringlet can be seen in the Encke Gap, crossing through center. This view looks toward the unlit side of the rings from about 18 degrees above the ring plane. Cassini took the image in visible blue light on November 4, 2006, at a distance of approximately 1 million miles (1.7 million kilometers) from Saturn. Image Credit: NASA/ JPL-Caltech/Space Science Institute

28

GROOVES ON BLUE

The Cassini spacecraft peers through Saturn's delicate, translucent inner C ring to see the diffuse blue limb of Saturn's atmosphere. This view looks toward the unilluminated side of the rings from about 20 degrees above the ring plane. It was created by combining images Cassini took using red, green, and blue spectral filters on April 25, 2008, at a distance of approximately 930,000 miles (1.5 million kilometers) from Saturn. Image Credit: NASA/JPL-Caltech/Space Science Institute

ATMOSPHERIC ILLUSION

Believe it or not, this extreme close-up of Saturn's swirling clouds was acquired from more than 621,370 miles (1 million kilometers) from the gas giant. The rings appear severely bent because of atmospheric refraction as they pass behind the planet. The dark region in the rings is the 2,980-mile-wide (4,800-kilometer-wide) Cassini Division. Cassini captured the image in visible light on June 25, 2005, at approximately 600,000 miles (1 million kilometers) from Saturn. Image Credit: NASA/JPL-Caltech/Space Science Institute

RAINBOW ON THE RINGS

The opposition effect, a brightness surge visible on Saturn's rings when the sun is directly behind the spacecraft, is captured here as a colorful halo. The rainbow is actually an artifact from how the image was produced. Cassini acquires color images by taking sequential exposures using red, green, and blue spectral filters, which are then composited together to form a color view. In this case the bright patch traveled across the rings between exposures (because Cassini was moving), creating three colorful spots from three separate moments. This view looks toward the sunlit side of the rings from about 9 degrees below the ring plane. Cassini acquired the images for this view on June 12, 2007, at approximately 325,000 miles (523,000 kilometers) from Saturn. Image Credit: NASA/JPL-Caltech/Space Science Institute

FLEEING THE SCENE

Saturn's moon Prometheus, having perturbed the planet's thin F ring, continues in its orbit. The gravity of potato-shaped Prometheus (53 miles, or 86 kilometers, across) periodically creates streamer-channels in the F ring, and the moon's handiwork is visible in the dark channels here. This view looks toward the northern, sunlit side of the rings from about 10 degrees above the ring plane. A star is visible through the rings near center-right of the image. Cassini took the image in visible light on June 1, 2010, at approximately 808,000 miles (1.3 million kilometers) from Saturn. Image Credit: NASA/JPL-Caltech/Space Science Institute

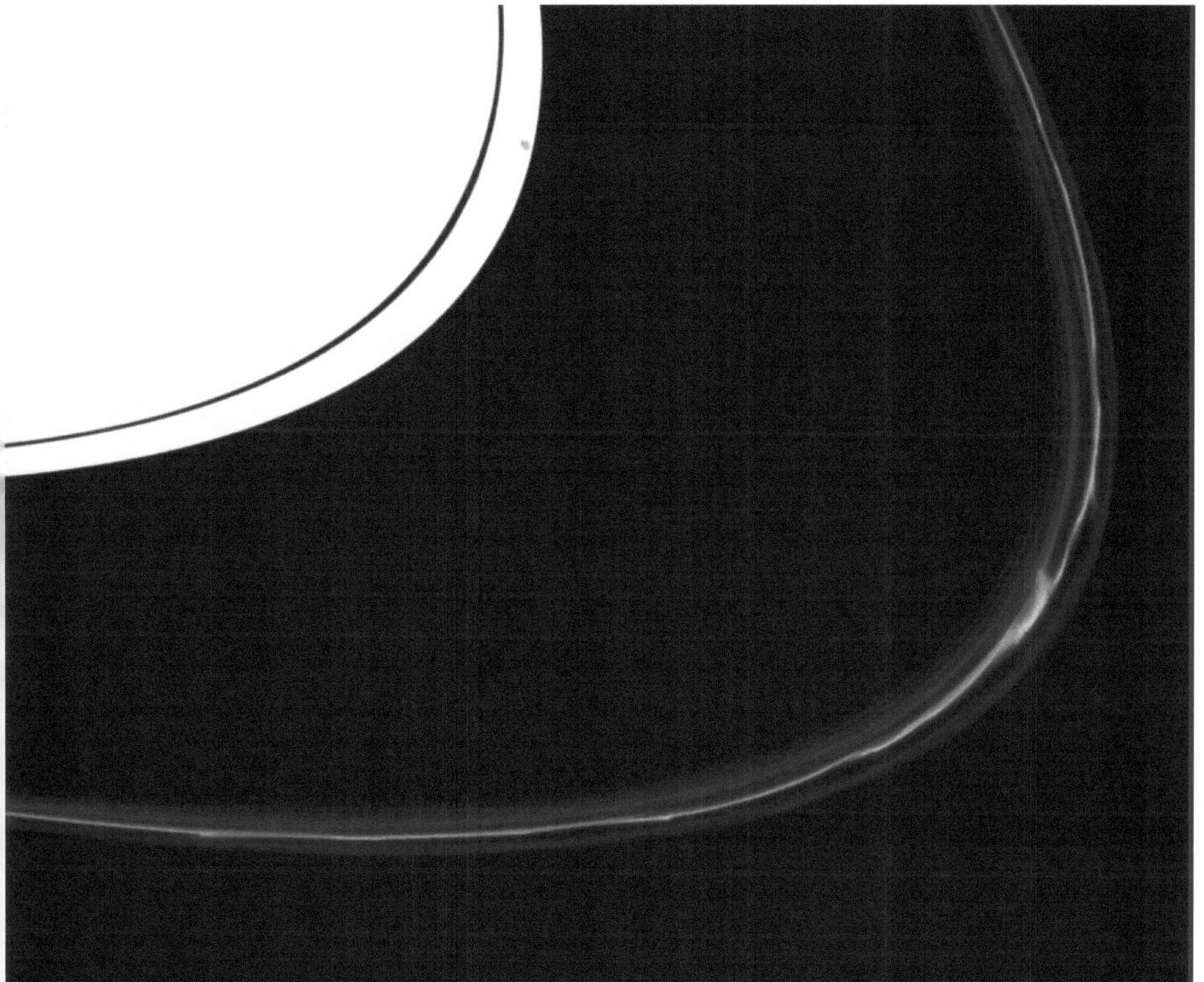

F RING ZOO

The F ring shows off a rich variety of phenomena in this image. Near the lower-right of the F ring, two "fans" of material radiate out of the main strand (or "core") of the ring. Kinks are apparent all along the core, and dark "channels" cut into the main strand — activity resulting from a recent interaction with the shepherd moon, Prometheus (which cannot be seen in this image). This view looks toward the sunlit side of the rings from about six degrees above the ring plane. Cassini took the image in visible light on December 25, 2012, at approximately 680,000 miles (1.1 million kilometers) from Saturn. Image Credit: NASA/JPL-Caltech/Space Science Institute

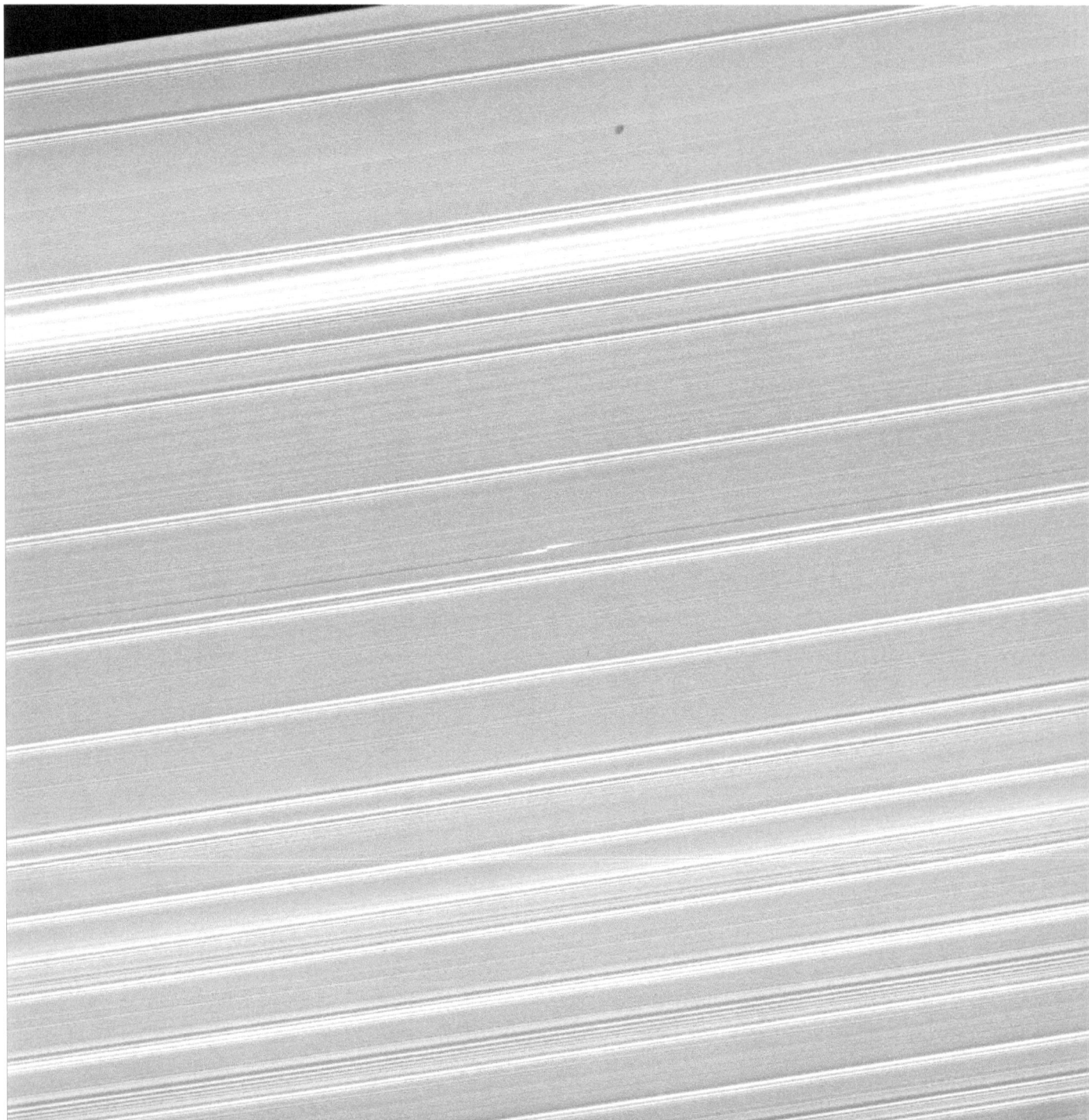

CHECKING IN ON BLERIOT

What appears as a pair of bright dashes at the center of this image is one of the features rings scientists have dubbed "propellers." This propeller, named Bleriot, marks the presence of a body that is much larger than the particles around it, yet too small to clear out a complete gap in the rings (as moons Pan and Daphnis have) and become a moon in its own right. The moonlet at the propeller's core is too small to see, but gravitational disturbances in the rings betray the moonlet's presence. Cassini took the image in visible light on January 9, 2017, at approximately 223,000 miles (359,000 kilometers) from Saturn. Image Credit: NASA/JPL-Caltech/Space Science Institute

PROPELLER BELTS OF SATURN

This view from NASA's Cassini spacecraft is the sharpest ever taken of belts of the features called propellers in the middle part of Saturn's A ring. The propellers are the small, bright features that look like double dashes, visible on both sides of the wave pattern that crosses the image diagonally from top to bottom. This image shows, for the first time, swarms of propellers of a wide range of sizes, putting the ones Cassini observed in its Saturn arrival images in context. Cassini took this image April 19, 2017, at approximately 80,000 miles (129,000 kilometers) from Saturn's center. Image Credit: NASA/JPL-Caltech/Space Science Institute

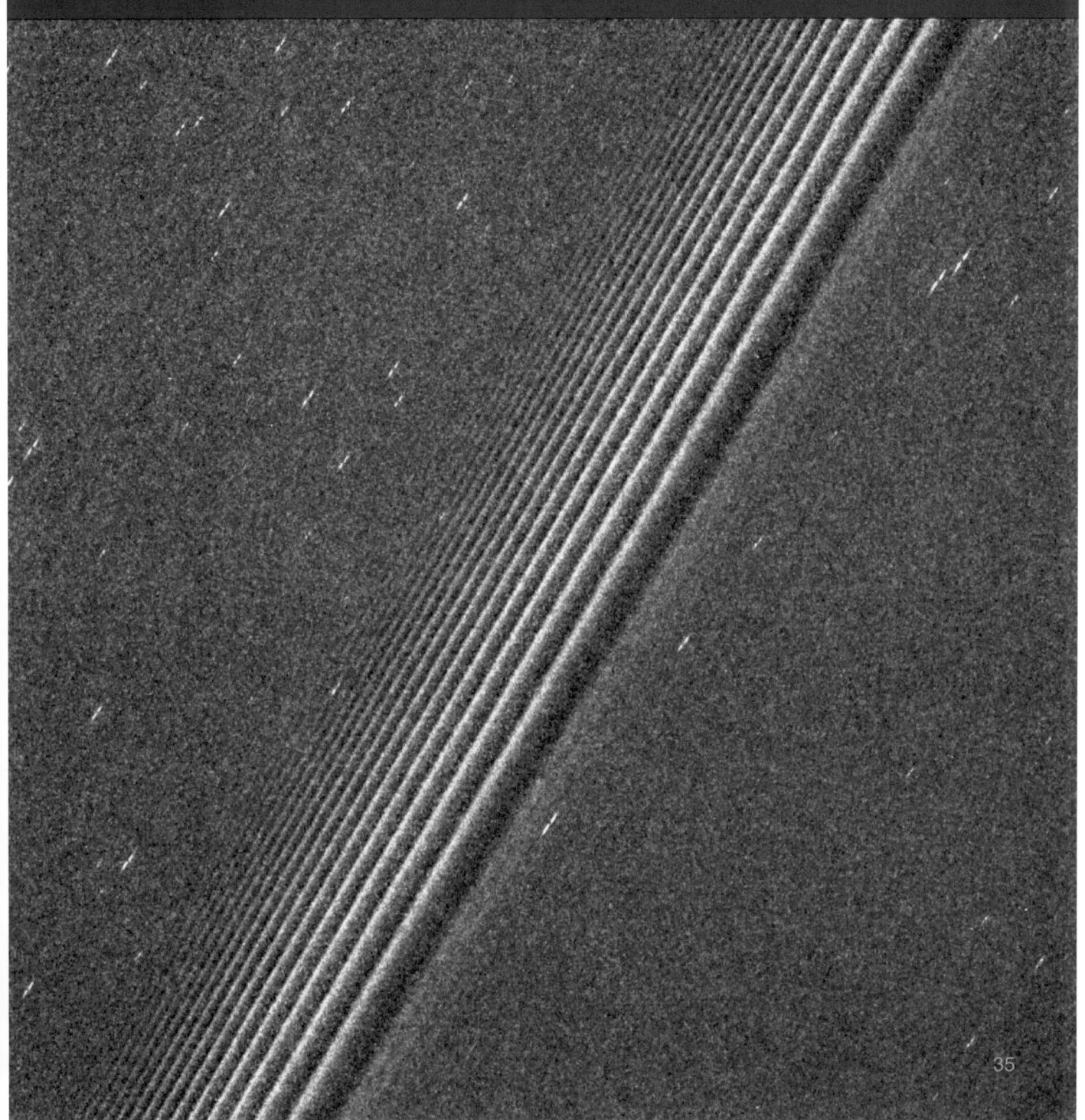

SAWTOOTH SHADOWS

Jagged-looking shadows stretch away from vertical structures of ring material created by the moon Daphnis in this image, taken as Saturn approached its August 2009 equinox, which occurs about only once every 15 Earth years. The illumination geometry in the few months before and after equinox causes out-of-plane structures to cast long shadows across the rings. Daphnis (5 miles, or 8 kilometers across) is a bright dot casting a thin shadow just left of center in the image. The moon has an inclined orbit, and it gravitationally perturbs the particles of the A ring forming the Keeler Gap's edge. Cassini took the image in visible light on June 26, 2009, at approximately 511,000 miles (823,000 kilometers) from Daphnis. Image Credit: NASA/JPL-Caltech/Space Science Institute

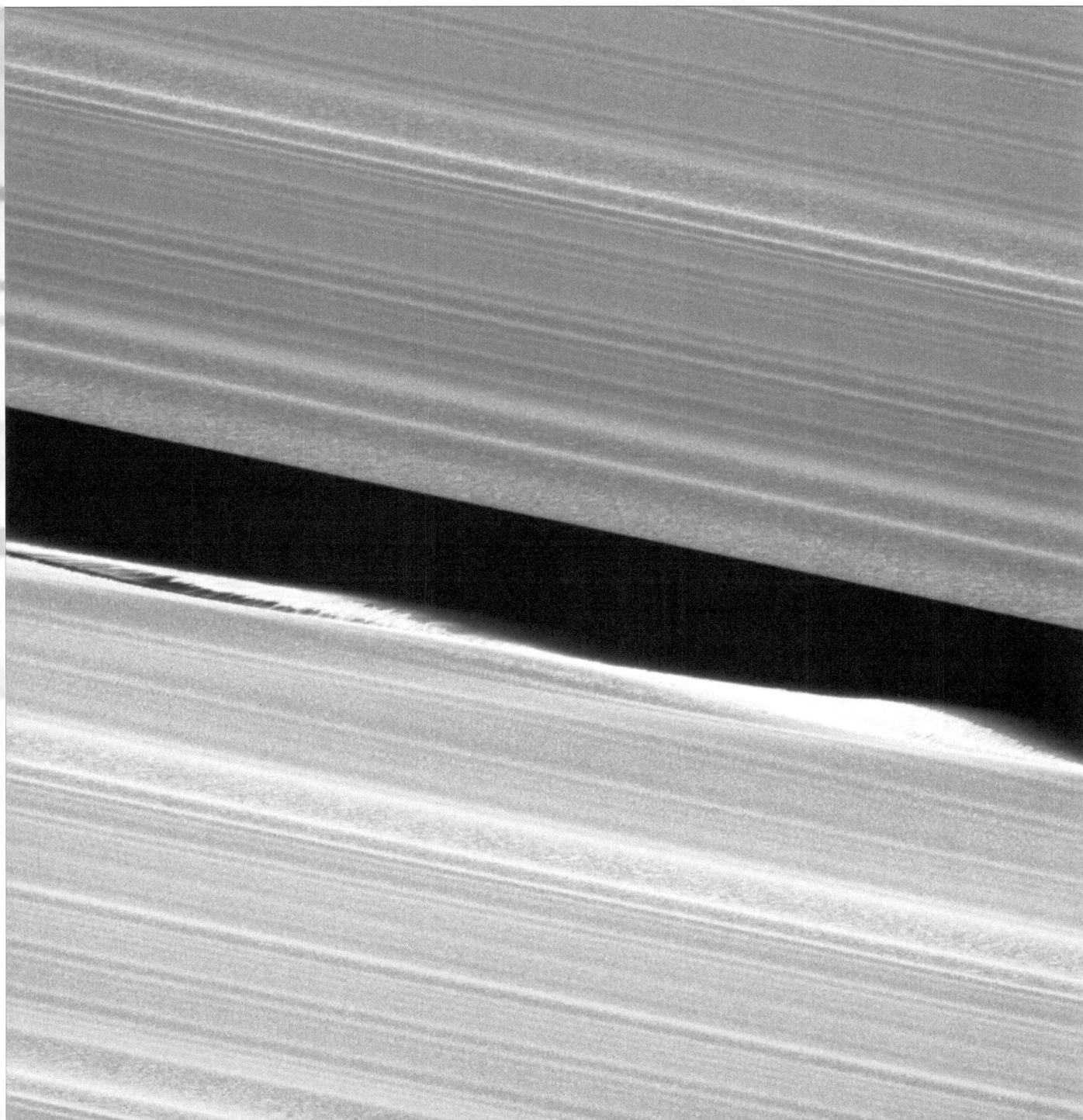

WAVING GOODBYE

This close-up view of the Keeler Gap, which is near the outer edge of Saturn's main rings, shows in great detail just how much the moon Daphnis affects the edges of the gap. Daphnis creates waves in the edges of the gap through its gravitational influence. Some clumping of ring particles can be seen in the perturbed edge. This view looks toward the sunlit side of the rings from about 3 degrees above the ring plane. Cassini acquired the image in visible light on January 16, 2017, at approximately 18,000 miles (30,000 kilometers) from Daphnis. Image Credit: NASA/JPL-Caltech/Space Science Institute

THE TALLEST PEAKS

Rising abruptly from the edge of Saturn's B ring are vertical structures casting long shadows on the ring in this image, taken by Cassini two weeks before the planet's August 2009 equinox. The vertical structures tower as high as 1.6 miles (2.5 kilometers) above the plane of the rings, which are generally only about about 30 feet (10 meters) thick. This view looks toward the southern, sunlit side of the rings from about 32 degrees below the ring plane. Cassini captured this image of a 750-mile-long (1,200-kilometer-long) section along the outer edge of the B ring in visible light on July 26, 2009. The view was acquired at a distance of approximately 209,000 miles (336,000 kilometers) from Saturn. Image Credit: NASA/JPL-Caltech/Space Science Institute

3

TITAN:
A MOON OBSCURED

Of the more than 150 known moons in our solar system, only one has a substantial atmosphere. Of all the worlds in the solar system — the moons, planets, dwarf planets, and small bodies — only one place besides Earth is known to have liquid lakes and seas on its surface. In both cases, that one place is Saturn's moon, Titan.

At 3,200 miles (5,150 kilometers) across, Titan is slightly larger than planet Mercury and is also perhaps the most mysterious moon in our solar system because (if for no other reason) the moon's surface is shrouded beneath an orange haze. This mammoth moon was one of the highest-priority scientific targets for the Cassini-Huygens mission. The spacecraft's Visual and Infrared Mapping Spectrometer and Cassini's RADAR instrument were included on the spacecraft largely to peek through Titan's veil. The same goes for the European Space Agency's Huygens probe, which journeyed to Saturn with Cassini, and whose sole purpose was to land on and study Titan, earning the honor of being the first probe to land in the outer solar system.

Cassini and Huygens found that if we set aside the difference in size, this fuzzy-looking and frigid world is more like Earth than any other place human spacecraft have explored. "Titan has so many of the processes we have on Earth but with exotic materials," said Jonathan Lunine, a Cassini interdisciplinary scientist from Cornell University. "It has methane-driven weather instead of water-driven weather, ice mountains instead of rock mountains, and organic dune particles instead of silica sand dune particles."

A human standing upon the surface of Titan would find the environment both familiar and alien. Like Earth, Titan has a primarily nitrogen atmosphere that's so dense that it's one of the most hospitable atmospheres in the solar system. "You just need oxygen to breathe and protection against the cold, but no pressure suit needed," Lunine said.

Titan's surface is coated in organic molecules that form in the upper atmosphere. Methane rain fills the rivers, lakes and seas, the largest of which are hundreds of feet deep and hundreds of miles wide. "You could hear the waves lapping onto the shore as you stand close to Ligeia Mare, one of the great methane seas of Titan," Lunine said. With the far shore beyond the horizon, it would be like standing on the shore of one of the U.S. Great Lakes, but with a few cosmetic alterations. "Remove the human-made structures, turn down the sun to twilight, tint the light red, and crank gravity down to lunar values," Lunine said. The relatively low gravity and dense atmosphere have a bizarre consequence. "The methane rain falls so slowly that a typical storm will pass over you and drift away before the rain arrives on your head — a methane sun-shower!"

Chapter Title Image:
Colorful Colossuses and Changing Hues

A giant of a moon appears before a giant of a planet undergoing seasonal changes in this natural-color view of Titan and Saturn from NASA's Cassini spacecraft. This view looks toward the northern, sunlit side of the rings from just above the ring plane. This mosaic combines six images — two each of red, green, and blue spectral filters — to create this natural-color view. The Cassini spacecraft captured the images on May 6, 2012, at approximately 483,000 miles (778,000 kilometers) from Titan. Image Credit: NASA/JPL-Caltech/Space Science Institute

VEILED WORLDS

Although Titan's and Saturn's faces seem similar here, appearances can be misleading. Saturn is a gas giant, covered in clouds, with no solid surface to speak of, while Titan's atmosphere is a blanket of dense haze — a photochemical smog — surrounding an icy, solid body. Saturn's atmosphere is mostly hydrogen and helium with clouds of water, ammonia, and ammonium hydrosulfide. Titan's atmosphere, however, is primarily nitrogen and methane, with occasional methane clouds. This view looks toward Saturn from the unilluminated side of the rings, 0.3 degrees below the ring plane. The image was taken in visible green light on May 22, 2015, at approximately 1.4 million miles (2.2 million kilometers) from Saturn. Image Credit: NASA/JPL-Caltech/Space Science Institute

MANY COLORS, MANY MOONS

Four moons huddle near Saturn's multi-hued disk in this natural-color view captured by Cassini on October 26, 2007, at approximately 920,000 miles (1.5 million kilometers) from Saturn and 1.7 million miles (2.7 million kilometers) from Titan. Giant Titan dominates the smaller moons in the scene. Beneath and left of Titan is Janus (113 miles, or 181 kilometers across). Mimas (247 miles, or 397 kilometers across) appears as a bright dot close to the planet and beneath the rings. Prometheus (63 miles, or 102 kilometers across) is a faint speck hugging the rings between the two small moons. This view looks toward the unilluminated side of the rings from less than a degree above the ring plane. Image Credit: NASA/JPL/Space Science Institute

POLAR VORTEX
IN COLOR

The recently formed south polar vortex stands out in the color-swaddled atmosphere of Saturn's largest moon, Titan, in this view acquired by Cassini on July 25, 2012, at approximately 64,000 miles (103,000 kilometers) from Titan. To create this natural-color view, images taken using red, green, and blue spectral filters were combined. Image Credit: NASA/JPL-Caltech/Space Science Institute

A RING OF COLOR

Sunlight scatters through the periphery of Titan's atmosphere, forming a ring of color in this view captured by Cassini on June 6, 2012, at approximately 134,000 miles (216,000 kilometers) from Titan. The hazy moon's north polar hood can be seen at the top of this view, and a hint of the south polar vortex is visible at bottom. Cassini is looking at the Saturn-facing side of Titan, where north is up and rotated 9 degrees to the right. Images taken using red, green, and blue spectral filters were combined to create this natural-color view. Image Credit: NASA/JPL-Caltech/Space Science Institute

FIRE AND ICE

Saturn's largest and second largest moons, Titan and Rhea, appear stacked atop one another in this true-color scene from Cassini. Titan's north polar hood appears as a detached layer at the top right of the moon. This view looks toward the Saturn-facing side of Rhea, whose north in this image is up and rotated 35 degrees to the right. The images that were combined to create this view were taken using red, green, and blue spectral filters on June 16, 2011, at a distance of approximately 1.1 million miles (1.8 million kilometers) from Rhea and 1.5 million miles (2.5 million kilometers) from Titan. Image Credit: NASA/JPL-Caltech/Space Science Institute

PEEKING AT SATURN

Cassini peers around the hazy limb of Titan to spy the sunlit south pole of Saturn in the distance beyond. The thick, smog-like atmosphere of frigid Titan was a major source of interest for the Cassini mission. The images that were combined to create this view were taken using red, green, and blue spectral filters on December 26, 2005, at a distance of approximately 16,000 miles (26,000 kilometers) from Titan. Image Credit: NASA/JPL/Space Science Institute

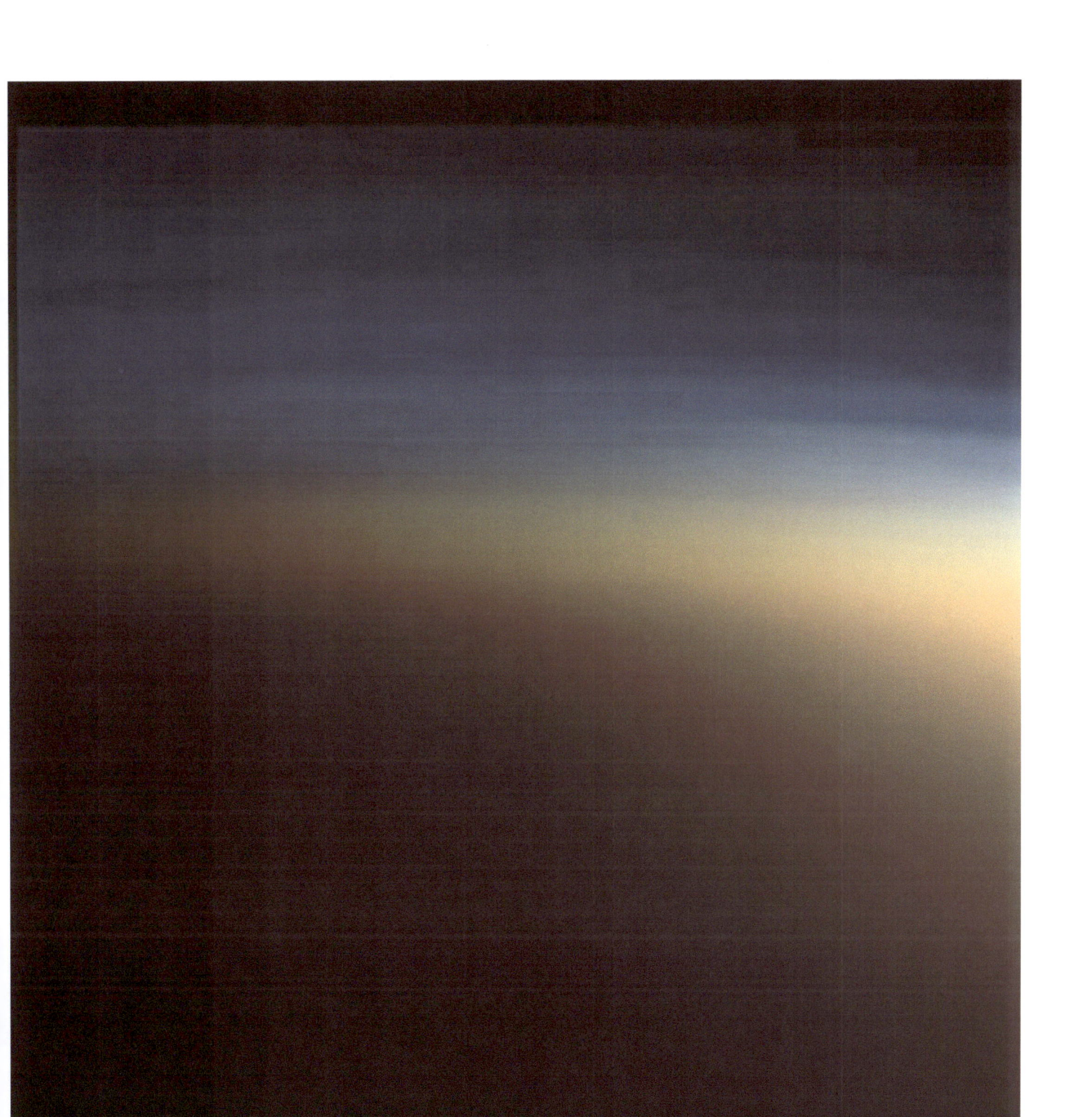

TITAN: COMPLEX "ANTI-GREENHOUSE"

This natural-color image shows Titan's upper atmosphere — an active place where solar ultraviolet light breaks apart methane molecules, and the byproducts combine to form compounds like ethane and acetylene. The haze preferentially scatters blue and ultraviolet wavelengths of light, making the atmosphere's complex layered structure more easily visible at the shorter wavelengths used in this image. The images that were combined to create this view were taken on March 31, 2005, using red, green, and blue spectral filters at approximately 5,900 miles (9,500 kilometers) from Titan.
Image Credit: NASA/JPL/Space Science Institute

REFLECTION OF SUNLIGHT OFF TITAN LAKE

This image shows the first flash of sunlight reflected off a hydrocarbon lake on Titan. The moon's north was shrouded in darkness for nearly 15 years (half of a Saturn year), but as the sun began illuminating the area again, Cassini captured this image during its 59th flyby of Titan on July 8, 2009, at about 120,000 miles (200,000 kilometers) from the moon. The observation confirmed the presence of liquid in the moon's northern hemisphere, where lakes are more numerous and larger than in the southern hemisphere. Cassini looked for glints in infrared wavelengths that were able to penetrate the moon's hazy atmosphere, which scatters and absorbs many wavelengths of light, including most visible wavelengths. Image Credit: NASA/JPL/University of Arizona/DLR

PEERING THROUGH
TITAN'S HAZE

NORTHERN SUMMER ON TITAN
(PAGE 49)

The Cassini spacecraft sees bright methane clouds drifting in the summer skies of Saturn's moon Titan, along with dark hydrocarbon lakes and seas clustered around the north pole. The Saturn system reached northern summer solstice on May 24, 2017, and when Cassini took this image on June 9, 2017, most of the surface in the moon's northern high latitudes was illuminated by the sun, unlike earlier in the mission. Cassini obtained the view at about 315,000 miles (507,000 kilometers) from Titan, using a spectral filter that preferentially admits wavelengths of near-infrared light. Image Credit: NASA/JPL-Caltech/Space Science Institute

PEERING THROUGH TITAN'S HAZE
(PREVIOUS PAGE)

This composite image shows an infrared view of Saturn's moon, Titan, from Cassini, acquired during the mission's "T-114" flyby on November 13, 2015. At visible wavelengths, this view would show only Titan's hazy atmosphere, but near-infrared wavelengths allow Cassini's vision to penetrate Titan's haze and reveal the moon's surface. The view features the parallel, dark, dune-filled regions named Fensal (to the north) and Aztlan (to the south), which form the shape of a sideways letter "H." Several places on the image are at higher resolution than elsewhere because they were acquired near closest approach. During this Titan flyby, the spacecraft's closest-approach altitude was 6,200 miles (10,000 kilometers). Image Credit: NASA/JPL/University of Arizona/ University of Idaho

VAST LIGEIA MARE IN FALSE COLOR

Ligeia Mare, shown here in a false-color Cassini image, is the second largest known body of liquid on Saturn's moon, Titan. It is filled with liquid hydrocarbons, such as ethane and methane, and is one of the many seas and lakes that bejewel Titan's north polar region. Cassini obtained the false-color mosaic of radar images between February 2006 and April 2007. Dark areas (low-radar return) are colored black, while bright regions (high-radar return) are colored yellow to white. Image Credit: NASA/JPL-Caltech/ASI/Cornell

ON THE SHORE OF LIGEIA MARE, TITAN

This 360-degree viewer represents what it might look like if you stood on the shore of Ligeia Mare, one of Titan's methane seas. The details in this view are artistic approximations because such details were too fine for the Cassini spacecraft's instruments to detect, but the large-scale topography was created using real observations from the spacecraft. Likewise, the presence and shape of clouds is based on imagery from the spacecraft, and the color of the sky and low angle of the sun are modeled on observed characteristics of Titan.

You are here.

INTERACTIVE 3.1 Select the image below to enter the viewer and explore the landscape in 360 degrees. Once inside the viewer, click and drag the image to look around. On a mobile device, you can enable navigation controlled with the movement of your device by toggling the icon in the upper left corner of the viewer.

SATURN THROUGH THE HAZE

Less than 20 minutes after Cassini's close approach to Titan on March 31, 2005, its cameras captured this view of Saturn through Titan's upper atmosphere. The northern part of Saturn's disk can be seen at the upper left with dark horizontal shadows cast upon Saturn by its rings. The diffuse bright regions of the image (below Saturn and at right) are light being scattered by haze in Titan's upper atmosphere. The image shows both how Titan's haze transmits light (from the attenuation of light from Saturn) and how the haze reflects light (from its brightness next to Saturn). The image was taken in visible light at 4,960 miles (7,980 kilometers) from Titan, and about 808,000 miles (1.3 million kilometers) from Saturn. Image Credit: NASA/JPL/Space Science Institute

VIEWS OF TITAN FROM DIFFERENT ALTITUDES

VIEWS OF TITAN FROM DIFFERENT ALTITUDES
(PREVIOUS PAGE)

This is a set of images acquired by the European Space Agency's Huygens probe in the four cardinal directions (north, south, east, west), at five different altitudes above Titan's surface. The images were taken as Huygens descended through Titan's atmosphere to land on the moon's surface on January 14, 2005. The probe was delivered to Titan by the Cassini spacecraft. Image Credit: ESA/NASA/JPL/University of Arizona

FIRST COLOR VIEW OF TITAN'S SURFACE

This image was taken on January 14, 2005, by the European Space Agency's Huygens probe upon its successful landing on Titan. This is a processed image that gives an indication of the actual color of Titan's surface. The two rock-like objects just below the middle of the image are about 6 inches (15 centimeters) (left) and about 1.5 inches (4 centimeters) (center) across respectively, and are about 33 inches (85 centimeters) from Huygens. Titan's surface was darker than expected and consisted of a mixture of water and hydrocarbon ice. Image Credit: NASA/JPL/ESA/ University of Arizona

TITAN AND DIONE

Saturn's fourth-largest moon, Dione (698 miles or 1,123 kilometers across), is visible through the haze of Saturn's largest moon, Titan (3,200 miles or 5,150 kilometers across), in this view of the two moons before the planet. Titan's north polar hood appears as a detached layer at the top of the moon. This view looks toward the northern, sunlit side of the rings from just above the ring plane, and north is up on the moons. The images that were combined to create this view were taken on May 21, 2011, using red, green, and blue spectral filters at approximately 1.4 million miles (2.3 million kilometers) from Titan and 2 million miles (3.2 million kilometers) from Dione. Image Credit: NASA/JPL-Caltech/Space Science Institute

OBSCURED BY RINGS

Saturn's rings obscure part of Titan's colorful visage in this Cassini image. Titan's south polar vortex is visible at the bottom. Visible near the top is the north polar hood, a cap of haze slightly darker than the rest of Titan's atmosphere. Saturn's shadow darkens the rings near the center of this view, but a sliver of illuminated Titan can be seen through the Cassini Division near the middle of that darkness. This view looks toward the northern, sunlit side of the rings from just above the ring plane. Cassini captured the images that were combined to create this natural-color view on May 16, 2012, using red, green, and blue spectral filters at approximately 1.9 million miles (3 million kilometers) from Titan. Image Credit: NASA/JPL-Caltech/Space Science Institute

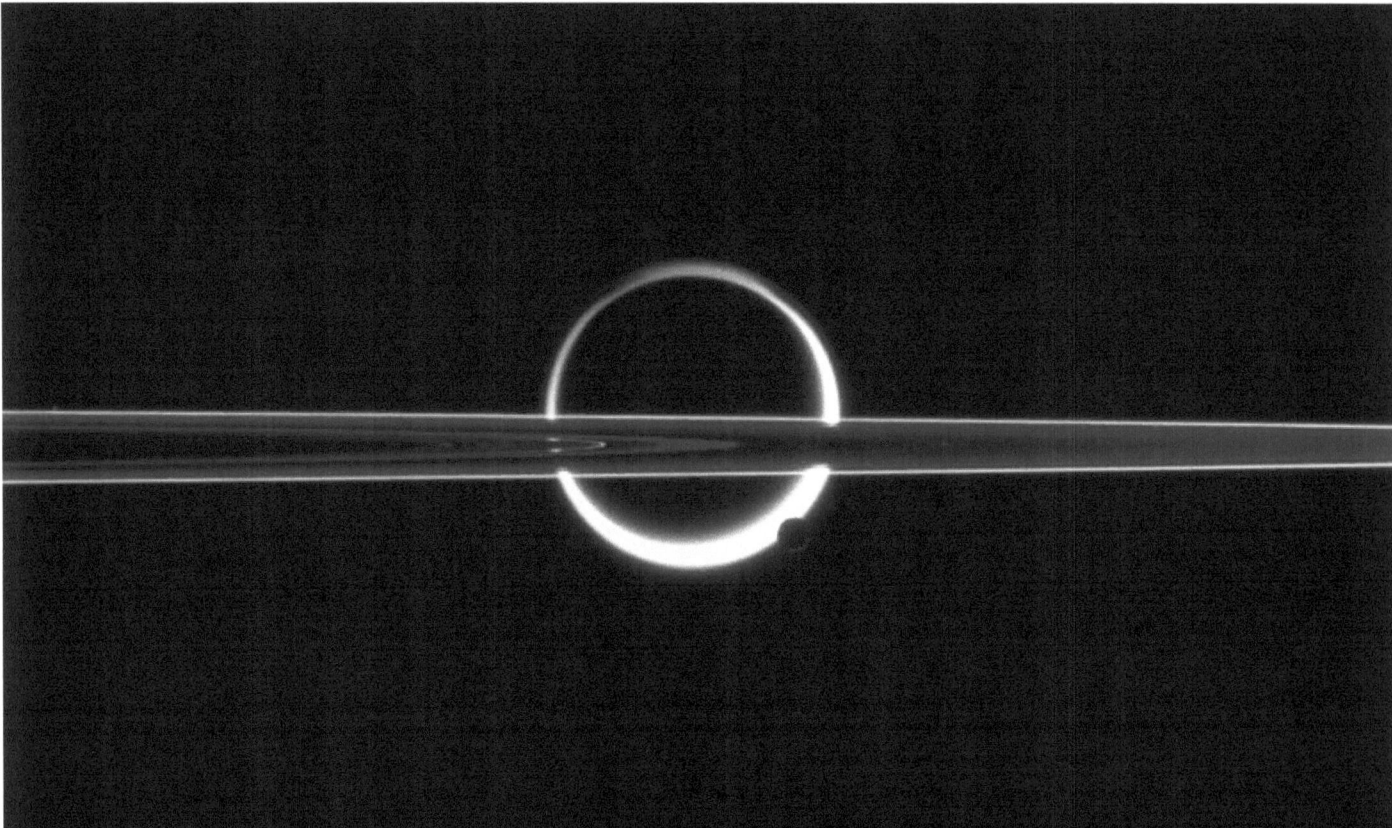

CANDLE IN THE DARK

Saturn's rings cut across an eerie scene ruled by Titan's luminous crescent and globe-encircling haze, broken by the small moon Enceladus, whose icy jets are dimly visible at its south pole. The scattered light around planet-sized Titan (3,200 miles, or 5,150 kilometers across) makes the moon's solid surface visible in silhouette. Enceladus (314 miles, or 505 kilometers across) enjoys far clearer skies than its giant sibling moon. This view shows the unlit side of Saturn's rings, and north is up. Cassini captured this image in visible red light on June 10, 2006, at approximately 2.4 million miles (3.9 million kilometers) from Enceladus and 3.3 million miles (5.3 million kilometers) from Titan. Image Credit: NASA/JPL/Space Science Institute

TITAN UP FRONT

The colorful globe of Saturn's largest moon, Titan, passes in front of the planet and its rings in this true-color snapshot from Cassini, which looks toward the northern, sunlit side of the rings from just above the ring plane. Titan's north polar hood appears as a detached layer at the top of the moon. Cassini captured the images that were combined to create this natural-color view on May 21, 2011, using red, green, and blue spectral filters at approximately 1.4 million miles (2.3 million kilometers) from Titan. Image Credit: NASA/JPL-Caltech/Space Science Institute

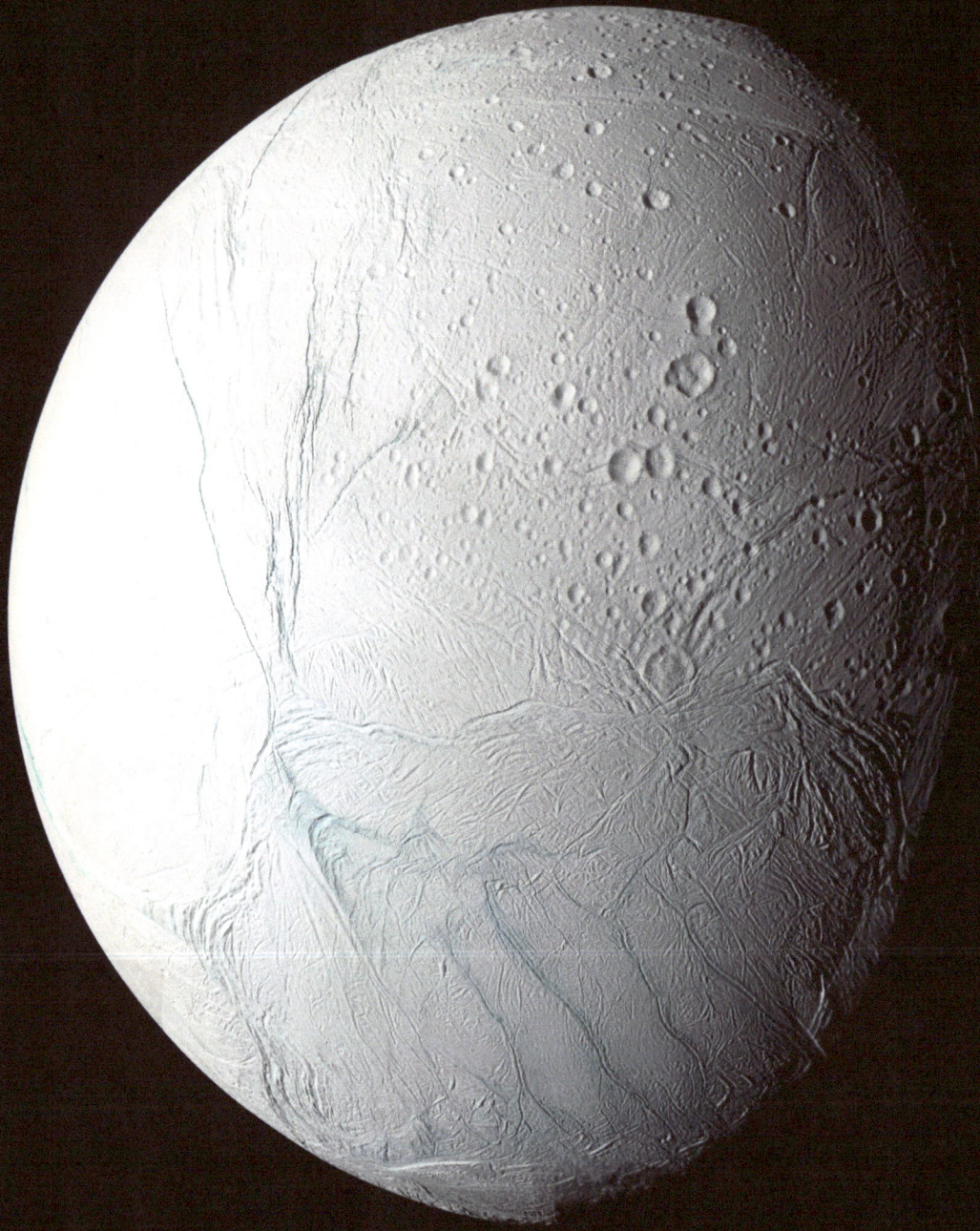

4

ENCELADUS:
THE RAREST OF PEARLS

Excluding Earth, Saturn's sixth largest moon, Enceladus, may be the most exceptional world in the solar system. Because Enceladus is so small — barely over 300 miles in diameter — and is nearly 10 times farther from the sun than Earth, it should be entirely frozen. But NASA's Cassini mission found that jets of water and icy particles spray from prominent fissures near the moon's south pole, producing a plume of ice and gas hundreds of miles high. That alone is extraordinary, but Cassini also revealed that the plume's source is a global ocean of salty water hidden beneath miles of ice. The ocean is venting directly into space through the fissures.

Most of the ice grains in the plume fall back to the surface of Enceladus, slowly softening the edges of the moon's craters and canyons the way snowfall on Earth covers footprints. This celestial snowfall makes Enceladus the whitest, most reflective world in the solar system. A fraction of the plume material escapes the moon's gravity to form Saturn's E ring — a ring entirely produced by a moon.

"We knew something was happening with this intriguing little ice moon — it looked like a winter wonderland of freshly fallen snow," said Bonnie Buratti, a senior research scientist at NASA's Jet Propulsion Laboratory. "When a hot spot was discovered at the south pole, and later a glorious plume of ice particles coming from that spot, we felt we had solved a great mystery. But there's still plenty of work to be done, characterizing that plume and how it forms and changes through time."

Hydrogen in the plume, along with tiny sand grains Cassini directly sampled and analyzed, provide convincing evidence that hydrothermal vents are spewing hot, mineral-laden water from below the seafloor into the Enceladus ocean. Such vents are one of the places scientists suspect life could have begun on Earth. Thanks to Cassini, Enceladus is now one of the prime targets in the quest to find environments beyond Earth where life might have evolved.

If you stood near the active south pole of this snow globe world, Saturn and its broad span of rings would nearly fill your view. From the half of the moon that faces Saturn, the ringed planet never sets or even seems to move because Enceladus always shows the same side to Saturn. But all is not still. The face of Saturn changes as the planet rotates on its axis every 10 hours. Another of Saturn's moons, Mimas, passes in front of Saturn every three days or so, just 33,000 miles from Enceladus. To Saturn's sides, the stars appear to move together from right to left in a cycle that repeats every 33 hours. If you turn away from Saturn completely, the blackness twinkles with glittering flecks of ice.

Chapter Title Image:
Zooming in on Enceladus

From afar, Enceladus exhibits a bizarre mixture of softened craters and complex, fractured terrains. This false-color mosaic was produced from 21 images Cassini acquired as the spacecraft swooped past the south pole of Saturn's moon, Enceladus, on July 14, 2005. The original images were taken at distances ranging from about 38,100 to 6,900 miles (61,300 to 11,100 kilometers) from Enceladus, and at wavelengths from the ultraviolet to the infrared portion of the spectrum. Image Credit: NASA/ JPL-Caltech/Space Science Institute

BULL'S-EYE MOONS

Since the two moons in this image are aligned and are at relatively similar distances from NASA's Cassini spacecraft, the image provides a good approximation of the relative sizes of Enceladus (313 miles or 504 kilometers across) and Tethys (660 miles or 1,062 kilometers across). This view looks toward the unilluminated side of the rings from 0.34 degrees below the ring plane. Cassini captured the image in red light on September 24, 2015. Image Credit: NASA/JPL-Caltech/Space Science Institute

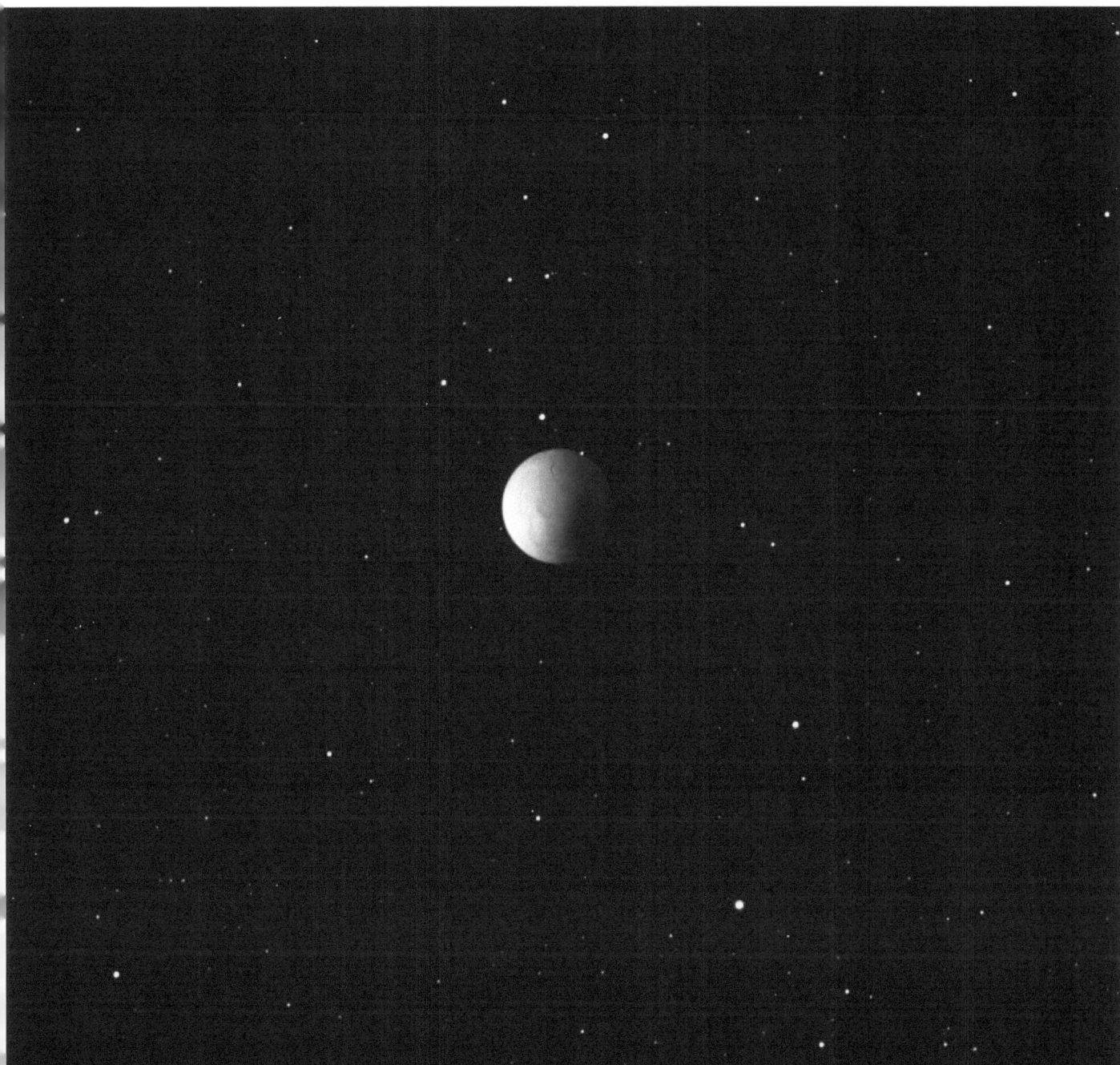

STARRY NIGHT

Numerous stars provide a serene background in this view of Enceladus captured by the Cassini spacecraft while the moon was in Saturn's shadow. The view looks up at Enceladus' south pole. Not visible from this angle are the icy moon's famed jets, which are aimed toward the spacecraft in this view. Cassini captured this image in visible light on October 9, 2008, at approximately 52,000 miles (83,000 kilometers) from Enceladus. Image Credit: NASA/JPL-Caltech/Space Science Institute

BRIGHT
ENCELADUS

Saturn's moon, Enceladus, reflects sunlight brightly in this image while the planet and its rings fill the background. Enceladus is the most reflective body in the solar system because fresh, white ice particles constantly coat its surface. In this view, north on Enceladus is up and rotated 21 degrees to the left. This view looks toward the northern, sunlit side of the rings from just above the ring plane. The image was taken in visible light on December 21, 2010, at approximately 63,000 miles (102,000 kilometers) from Enceladus. Image Credit: NASA/JPL-Caltech/Space Science Institute

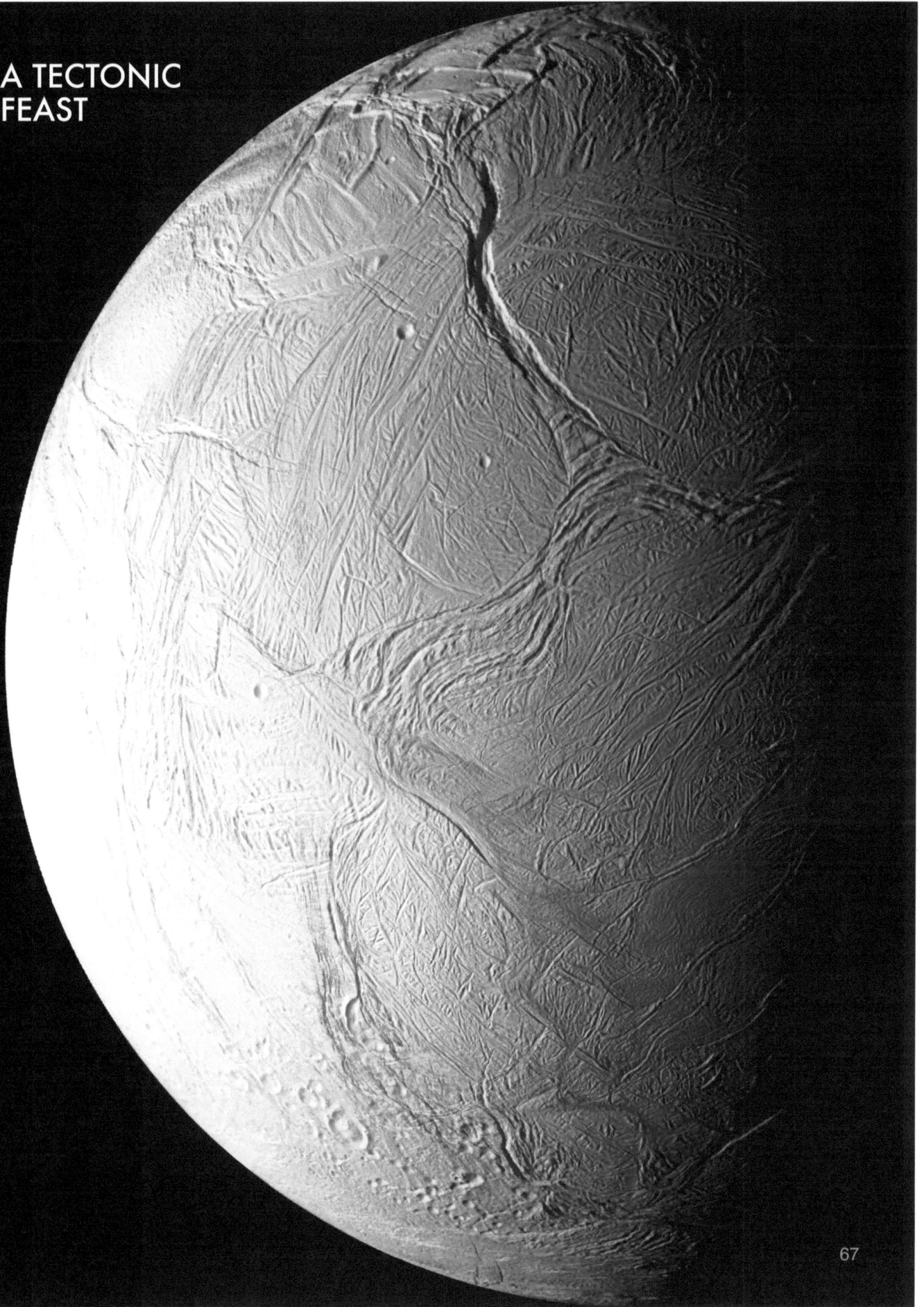

A TECTONIC
FEAST

A TECTONIC FEAST
(PREVIOUS PAGE)

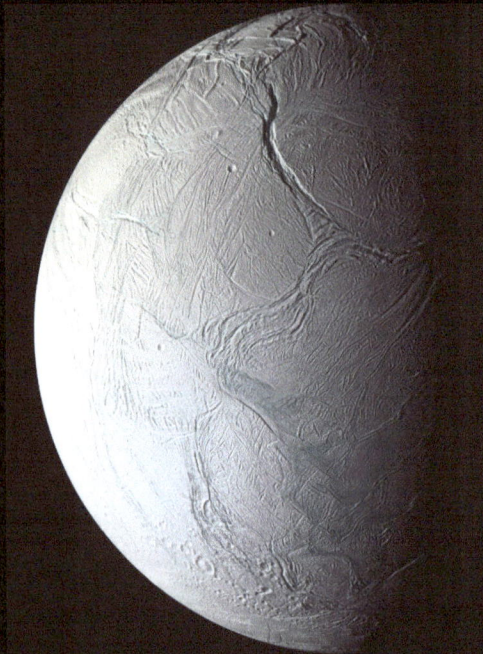

As Cassini receded from Enceladus just after coming within 15.6 miles (25 kilometers) of the moon's surface, the spacecraft captured 28 images used to create this false-color mosaic on October 9, 2008, at distances ranging from 18,750 to 30,000 miles (30,000 to 48,000 kilometers). Craters are rare in the southern region of the moon's Saturn-facing hemisphere. Instead are fractures, folds, and ridges — hallmarks of remarkable tectonic activity for a relatively small world. Image Credit: NASA/JPL-Caltech/ Space Science Institute

INTERACTIVE 4.1 **Enceladus**

This interactive visualization of Saturn's moon, Enceladus, was created from images captured by the Cassini spacecraft. Multiple images were stitched together to create a mosaic of the moon's surface, and that mosaic was then projected onto the model.

Select the image above to interact with the model.

FOUNTAINS OF ENCELADUS

This greatly enhanced and colorized image shows the enormous extent of the fainter, larger-scale component of Enceladus' plume, which is produced by water jets spraying through fissures at the south pole. Cassini captured this image on November 27, 2005. Image Credit: NASA/JPL-Caltech/ Space Science Institute

BEAUTIFUL PLUMAGE

Here sunlight reflected off Saturn illuminates Enceladus. North on Enceladus is up and rotated 45 degrees to the right. Cassini took the image in visible light on January 18, 2013, at approximately 483,000 miles (777,000 kilometers) from Enceladus. Image Credit: NASA/JPL-Caltech/Space Science Institute

BURSTING AT THE SEAMS:
THE WATER-JET BASIN OF ENCELADUS

Cassini captured this view of the south polar region of Enceladus on November 30, 2010. Jets of icy particles erupt continuously from fractures in the moon's crust. In this view, the plume created by the jets is backlit by the sun and falls partially within the shadow cast by Enceladus. Image Credit: NASA/JPL-Caltech/Space Science Institute

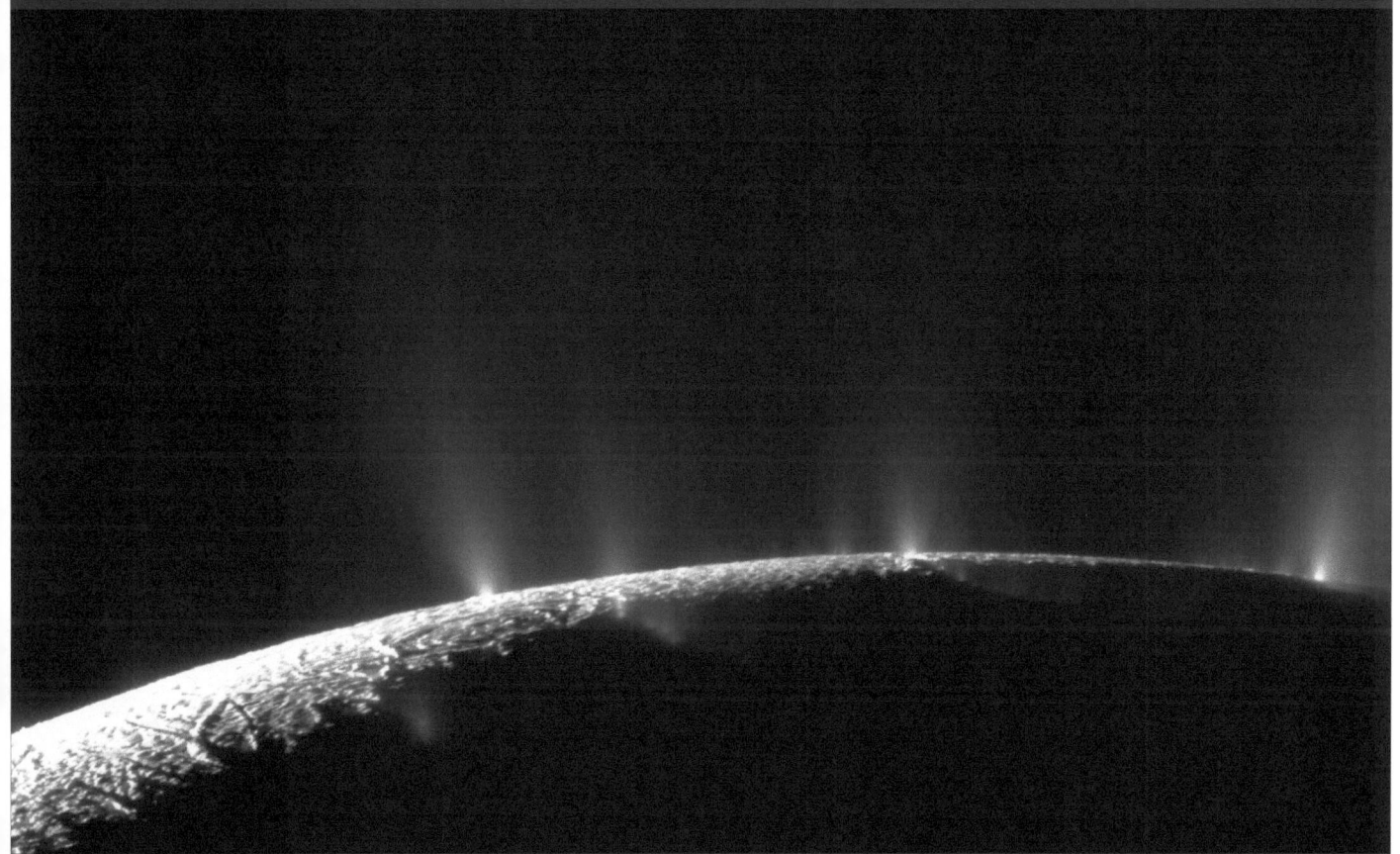

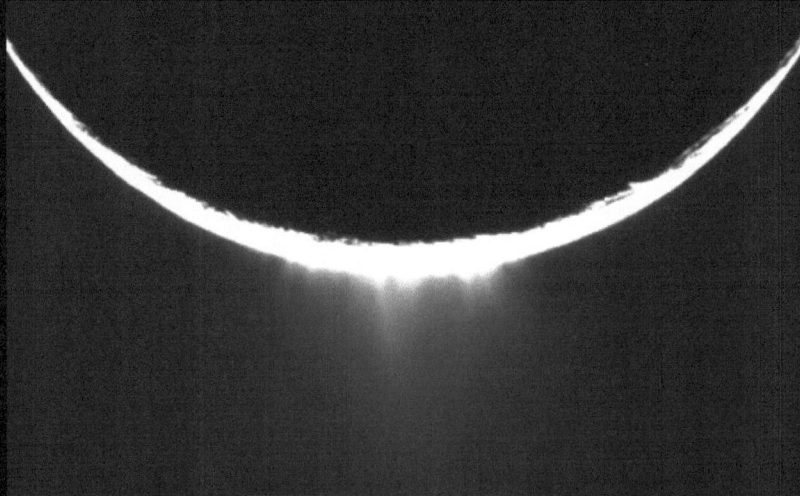

ENCELADUS PLUME MOVIE

Jets of icy particles burst from Saturn's moon, Enceladus, in this brief movie sequence of four images taken on November 27, 2005. The sensational discovery of active eruptions on a third outer solar system body (Io and Triton are the others) is one of the great highlights of the Cassini mission. These images were obtained at distances between about 89,700 and 92,900 miles (about 144,500 and 149,500 kilometers) from Enceladus. Image Credit: NASA/JPL-Caltech/Space Science Institute

Select the image above to view the sequence as a movie.

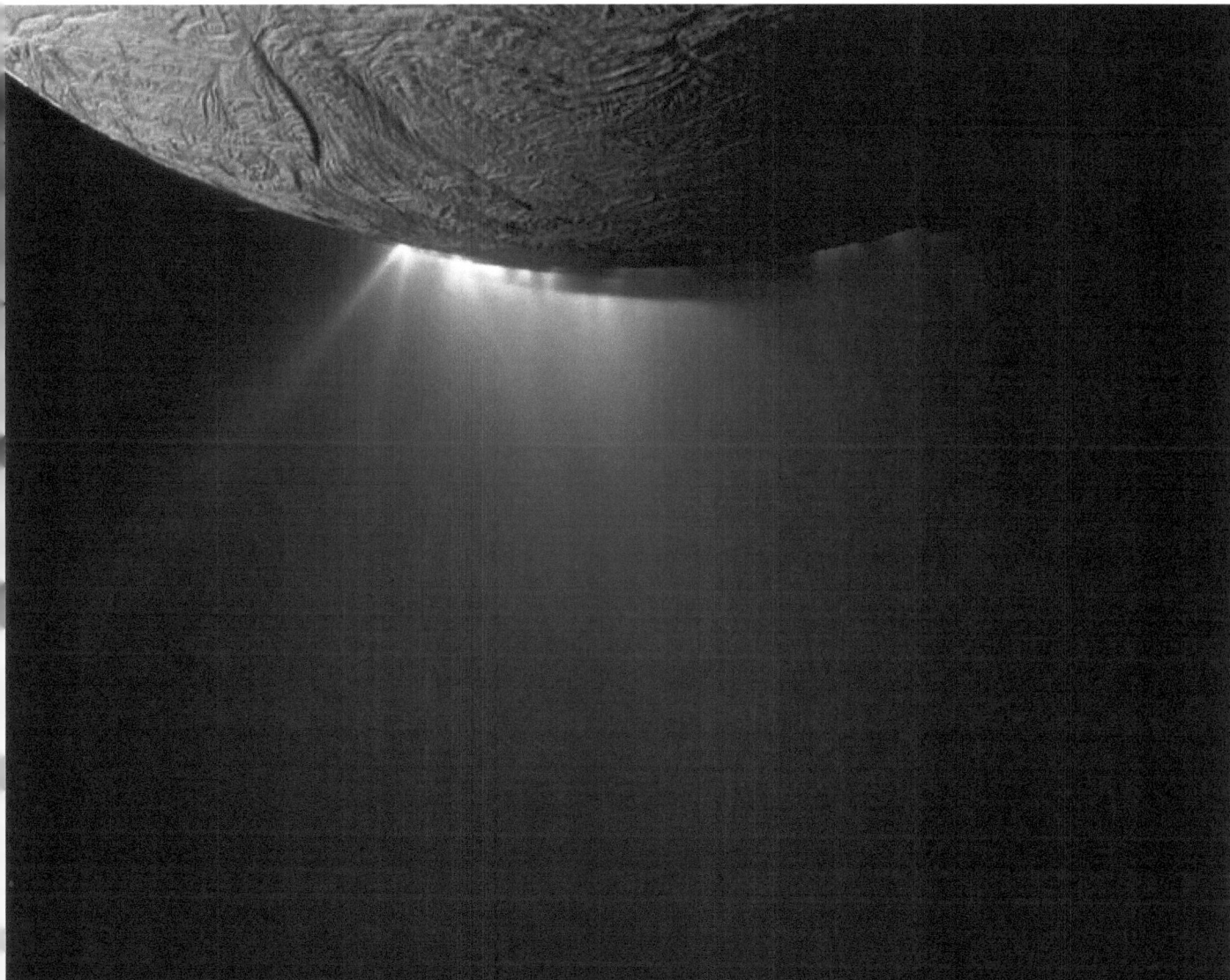

JETS AND SHADOWS

Cassini acquired this image on November 30, 2010, 1.4 years after Saturn's southern autumnal equinox, during a survey of the water-jet basin at the south pole of Enceladus. At the moment the image was captured, Cassini was essentially in the moon's equatorial plane, looking across the moon's south pole. The shadow of the body of Enceladus on the lower portions of the jets is clearly visible. Image Credit: NASA/JPL-Caltech/Space Science Institute

ENCELADUS FLYBY – SKEET SHOOT

This image was the eighth of several "skeet shoot" images Cassini captured during its October 31, 2008, flyby of Saturn's moon, Enceladus. The great fissure running across the image from left to right is Damascus Sulcus — one of the fractures through which the moon's salty ocean sprays into space. Cassini took the image at approximately 3,500 miles (5,600 kilometers) from Enceladus. Image Credit: NASA/JPL-Caltech/Space Science Institute

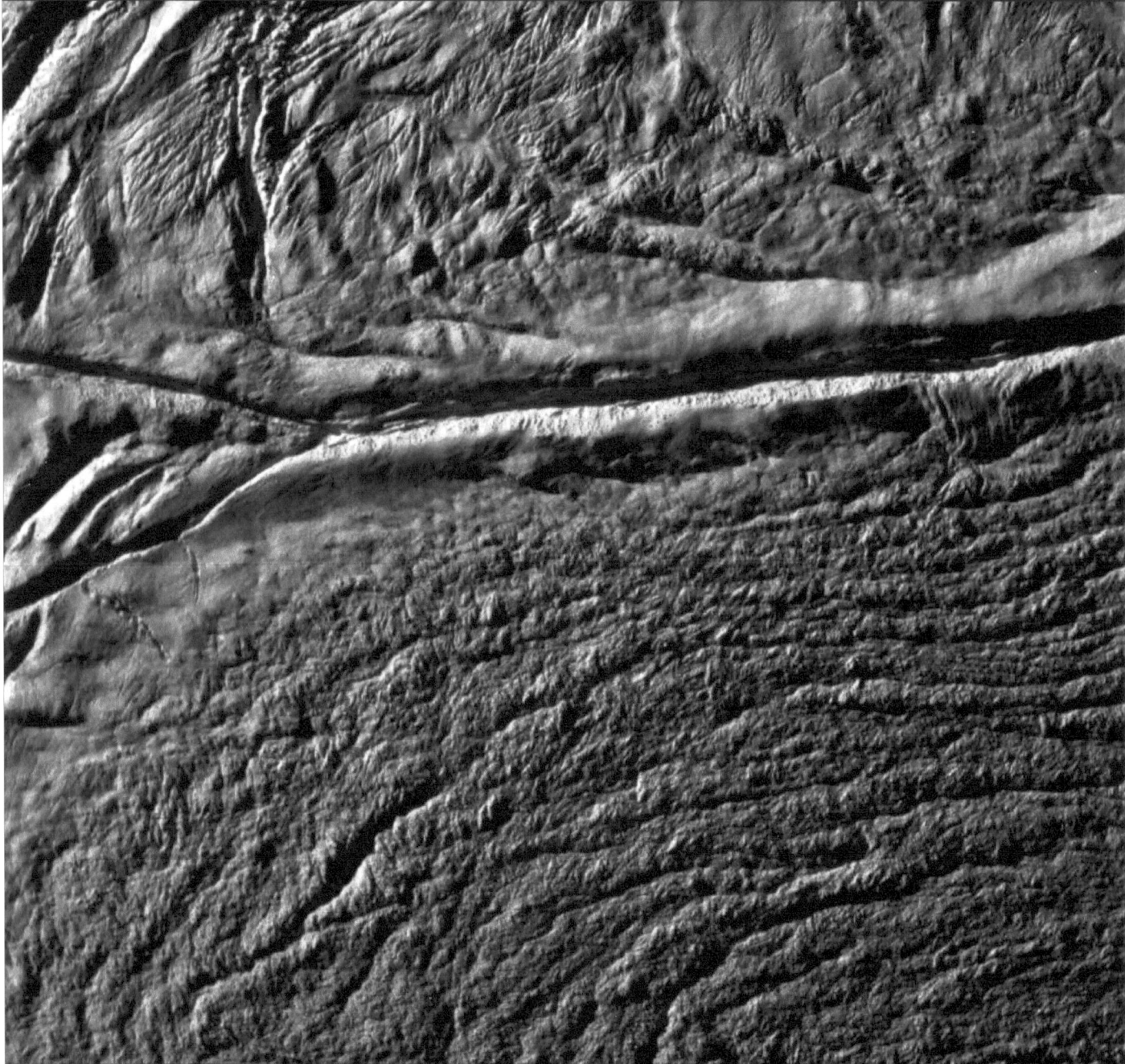

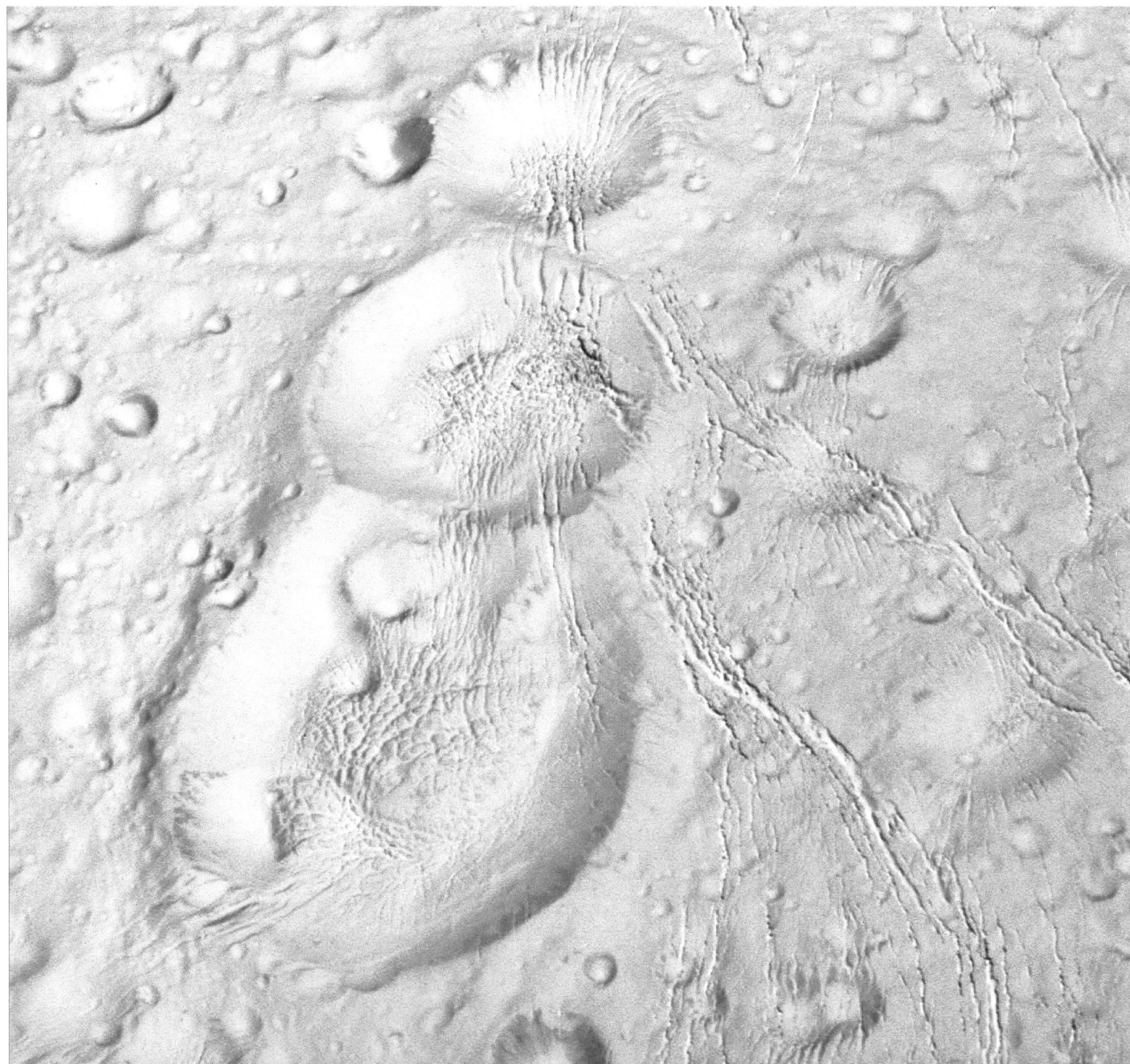

SATURNIAN SNOWMAN

NASA's Cassini spacecraft spied this tight trio of craters as it approached Saturn's icy moon, Enceladus, for a close flyby on October 14, 2015. The craters, located at high northern latitudes, are sliced through by thin fractures — part of a network of similar cracks that wrap around the snow-white moon. Cassini captured the image in visible light at approximately 6,000 miles (10,000 kilometers) from Enceladus. Image Credit: NASA/JPL-Caltech/ Space Science Institute

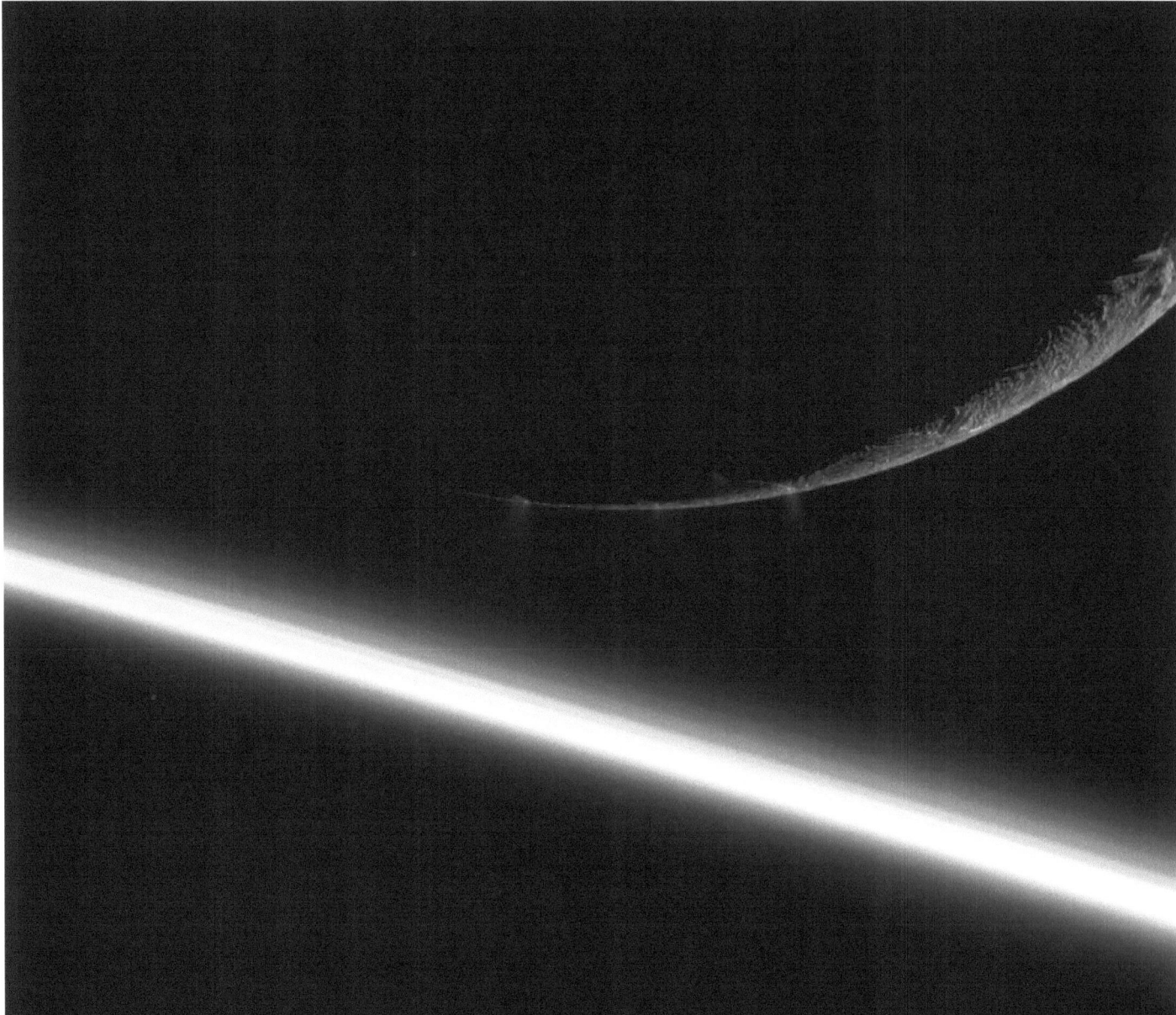

LOOMING ENCELADUS

Bottom left in this view is the night side of Saturn. Sunlight scatters through the planet's atmosphere and forms the bright diagonal line running from the left to bottom right of the image. Cassini captured the image during the August 13, 2010, flyby of Enceladus, which is top right and closer to the spacecraft than the planet is. The famous jets are faintly visible here erupting from the fractures that cross the south polar region of the moon. Cassini took the image in visible light at approximately 37,000 miles (59,000 kilometers) from Enceladus. Image Credit: NASA/ JPL-Caltech/Space Science Institute

FOCUS ON ENCELADUS

Ring shadows line the face of distant Saturn, providing an exquisite backdrop for the brilliant, white sphere of Enceladus. North is up. Cassini captured this image in green light on June 28, 2007, at approximately 181,000 miles (291,000 kilometers) from Enceladus. Image Credit: NASA/JPL-Caltech/Space Science Institute

LIGHT FROM MANY PATHS

The light Cassini captured in this view has traveled many paths. The unlit side of the rings glows with scattered sunlight as two moons circle giant Saturn. At left, the moon Mimas presents its dark side. On the far side of the rings, the moon Enceladus is lit by Saturnshine, or sunlight reflected from the planet. Saturn, in turn, is faintly lit in the south by light reflecting off the rings. Saturn's shadow darkens the rings, tapering off toward the left side of this view. The image was taken in visible light on June 11, 2006, at approximately 2.5 million miles (3.9 million kilometers) from Mimas, 2.7 million miles (4.3 kilometers) from Enceladus, and 2.6 million miles (4.1 million kilometers) from Saturn. Image Credit: NASA/JPL-Caltech/Space Science Institute

A PEARL AT DUSK

Enceladus hangs like a single bright pearl against the golden-brown canvas of Saturn and its icy rings. Visible on Saturn is the planet's terminator — the region where daylight gives way to dusk. Above, the rings throw thin shadows onto the planet. When approximately 100,000 miles (200,000 kilometers) from Enceladus, Cassini captured images using red, green, and blue spectral filters on January 17, 2006, and the images were combined to create this natural-color view. Image Credit: NASA/JPL-Caltech/Space Science Institute

THE TALE CONTINUES...

Cassini captured this view from just above the ring plane on March 12, 2012. From here, the brightly reflective moon Enceladus is before the rings, while the moon Titan is faintly visible in the background beyond the rings. Cassini captured the image in green light at approximately 600,000 miles (1 million kilometers) from Enceladus. Image Credit: NASA/JPL-Caltech/ Space Science Institute

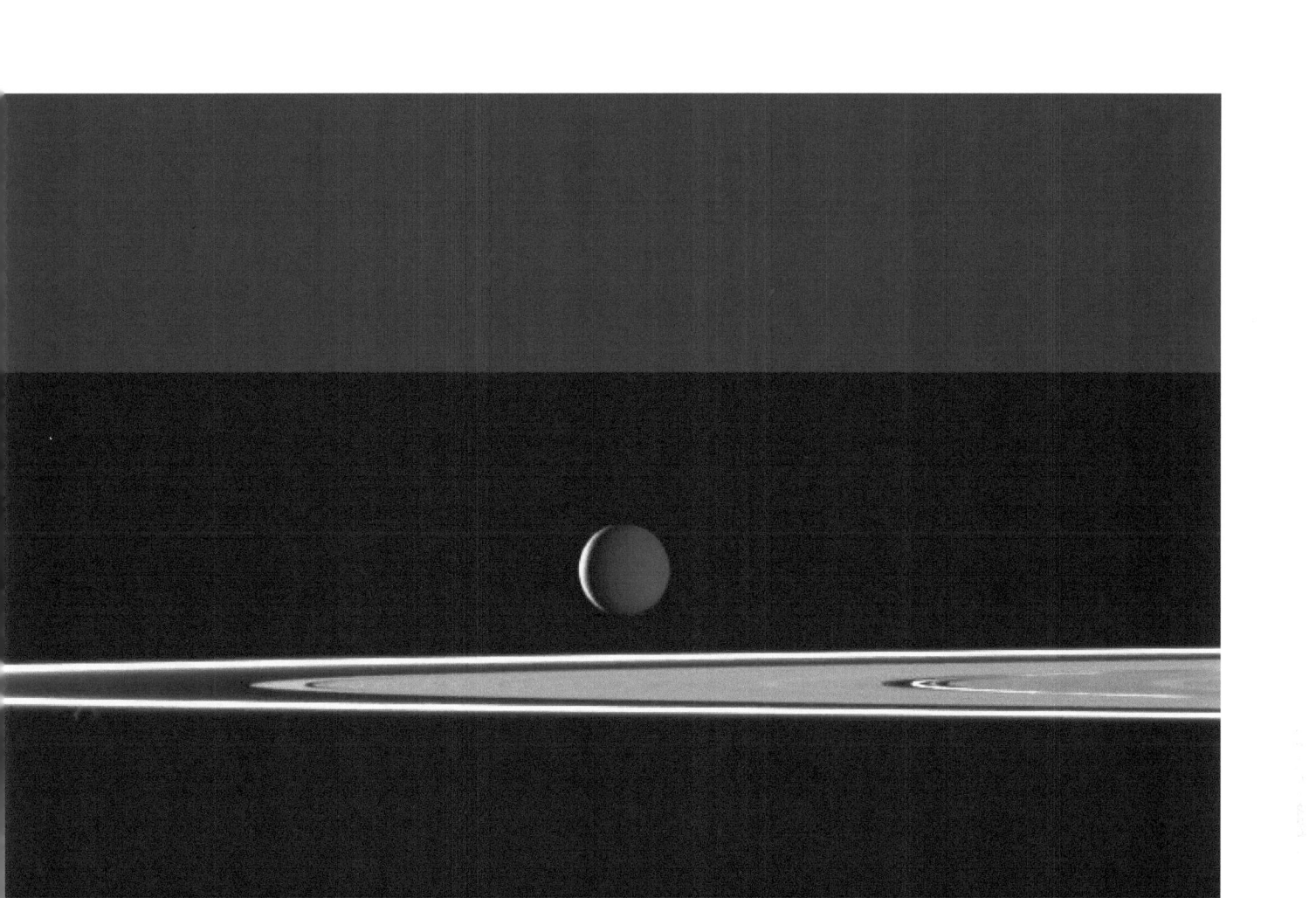

DUAL ILLUMINATED ENCELADUS

Sunlight directly illuminates a thin crescent of Enceladus on the moon's leading hemisphere, while sunlight reflected off Saturn dimly lights most of the moon in this image acquired in visible light on February 11, 2010, at approximately 932,000 miles (1.5 million kilometers) from Enceladus. This view looks toward the northern, sunlit side of the rings from just above the ring plane. Enceladus is more distant than the rings in this view while the small moon, Pandora, visible on the left of the image, is on the side of the rings nearest Cassini and illuminated by sunlight and Saturnshine. Image Credit: NASA/JPL-Caltech/Space Science Institute

HIGH-PHASE PLUMES

Sunlight brightly illuminates terrain on the left of Enceladus, while light reflected off Saturn illuminates the rest of the moon more dimly. Cassini took this image in visible light on October 13, 2009, at approximately 268,000 miles (431,000 kilometers) from Enceladus, and at a phase, or sun-Enceladus-spacecraft, angle of 159 degrees so that sunlight would reveal the backlit plume. North is up. Background stars, elongated by the movement of the spacecraft during the exposure, are also visible. Image Credit: NASA/JPL-Caltech/Space Science Institute

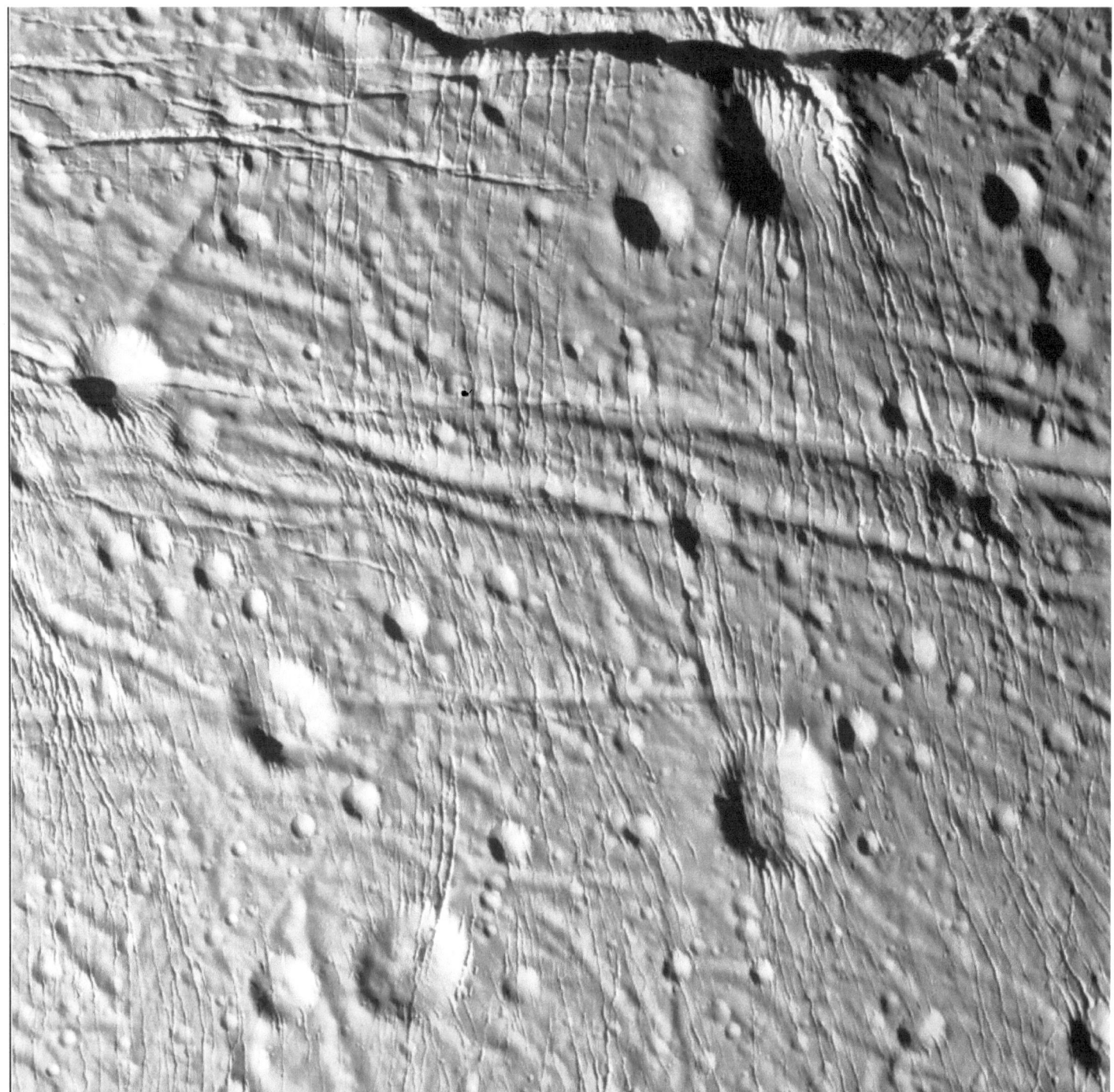

SLICED-UP CRATERS

During its very close flyby of Enceladus on March 9, 2005, Cassini took high-resolution images of the icy moon. This scene is an icy landscape that has been scored by tectonic forces. Many of the craters in this terrain have been heavily modified, such as the 6-mile-wide (10-kilometer-wide) crater near the upper right that has prominent north-south fracturing along its northeastern slope. The image has been rotated so that north on Enceladus is up. The image was taken in visible light from a distance of about 7,400 miles (11,900 kilometers) from Enceladus, and at a Sun-Enceladus-spacecraft, or phase, angle of 44 degrees. Image Credit: NASA/JPL-Caltech/Space Science Institute

ELEVATED VIEW OF ENCELADUS' SOUTH POLE

This dramatic view looks across the region of Enceladus' water-jet basin and down on the ends of the Baghdad and Damascus fractures that face Saturn. The segments of the fractures seen here are among the most active and warmest in the whole region. But because of the spacecraft's position and viewing angle, the jets are projected against the bright surface as opposed to black sky. As a result, the jets appear fuzzy, or indistinct. Cassini took the image through the clear filter on August 13, 2010, with a sun-Enceladus-spacecraft, or phase, angle of about 151 degrees. Image Credit: NASA/JPL-Caltech/Space Science Institute

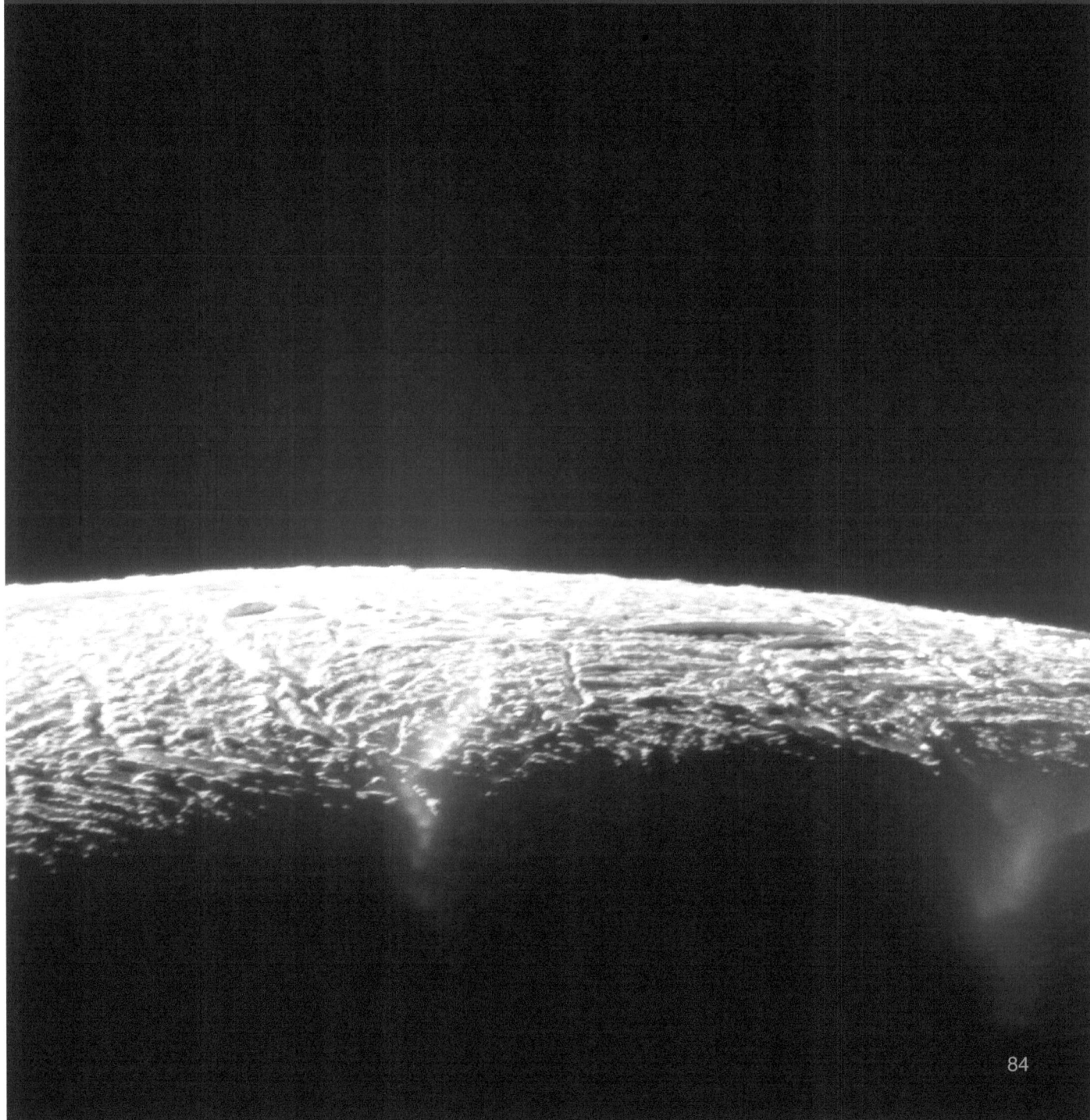

ENCELADUS IN FALSE COLOR

The surface of Enceladus shows a range of crater ages, as well as regions with very few discernible craters. In this view, the sun illuminates Enceladus from the lower left. Particularly visible are the four enormous fractures in the south polar region (seen here at the lower right) through which jets of icy particles erupt continuously. This false-color view is a composite of individual frames Cassini obtained using filters sensitive to ultraviolet, green, and infrared light on July 14, 2005, as the spacecraft approached Enceladus for an extremely close flyby. The view was enhanced to accentuate subtle color differences and fine-scale surface features. Cassini took the images from about 69,700 miles (112,100 kilometers) from Enceladus. Image Credit: NASA/JPL-Caltech/Space Science Institute

OTHER MOONS:
A MENAGERIE OF ICY WORLDS

If you left Earth and traveled toward Jupiter but kept going until you were roughly twice Jupiter's distance, you'd be in Saturn's realm. There, the inner planets are so distant that they could be mistaken for stars. If you waited 30 years, Saturn would inevitably lumber by, but Saturn does not wander this darkness alone.

Though Saturn's moons, Titan and Enceladus, receive a great deal of attention (because they may harbor environments where some form of life could survive), dozens of other worlds also swarm around the ringed planet. "Saturn has a treasure trove of moons!" said Amanda Hendrix, a planetary scientist at the Planetary Science Institute and a member of Cassini's ultraviolet spectrograph team who studies icy satellite data. "Each one is wonderful and unique," she said.

Some of Saturn's moons resemble potatoes or overstuffed ravioli. A couple of Saturn's moons may have had ring systems of their own at one time, and at least one moon orbits within a gap in Saturn's rings, gravitationaly kicking up ring particles as it goes. Another has colorful streaks of unknown origin. Saturn's moons range in size from about the size of a sports arena up to even larger than Mercury. They vary in color, texture, and composition — in a way, each Saturn moon is its own character. "Iapetus is one of my favorites," Hendrix said. "With one very dark hemisphere and one quite bright, it is such an oddball. And that equatorial ridge gives Iapetus a walnut-like appearance."

Not all of these worlds formed around Saturn. A few were more likely space rocks wandering through the neighborhood when Saturn's gravity invited them to the party, where they've remained ever since. Their names, too, have various origins, including characters from Greco-Roman, Inuit, Norse, and Gallic mythologies.

Before Cassini visited Saturn, the best images of many of the planet's moons showed only blurry or pixelated specks. Cassini has since transformed the many worlds of the Saturnian system into real places in their own right, but Hendrix says they would be challenging places for humans to visit. "They have no atmospheres to speak of, and they're very small, so there's very little gravity. And it's cold out there!" Hendrix said. "But those disadvantages aside, exploring these moons would be really cool. How about hiking the icy cliffs of Dione?! And checking out up close those mysterious red streaks on Tethys?"

Chapter Title Image:
The Other Side of Iapetus

This false-color mosaic shows the entire hemisphere of Saturn's moon Iapetus (912 miles, or 1,468 kilometers, across) visible from Cassini on September 10, 2007, at about 45,000 miles (73,000 kilometers) from Iapetus. Visible for the first time in detail are geologic structures that mark the trailing hemisphere, which appears heavily cratered, particularly in the north and south polar regions. The most prominent topographic feature in this view, located in the bottom half of the mosaic, is a 280-mile-wide (450-kilometer-wide) impact basin, one of at least nine such large basins on Iapetus. The basin overlaps an older, similar-sized impact basin to its southeast. Image Credit: NASA/JPL-Caltech/Space Science Institute

TRIPLE CRESCENTS

Three moons are shown here: Titan (3,200 miles or 5,150 kilometers across), Mimas (246 miles or 396 kilometers across), and Rhea (949 miles or 1,527 kilometers across). Titan, the largest moon in this image, appears fuzzy because we only see its cloud layers. Because Titan's atmosphere refracts light around the moon, its crescent "wraps" just a little further around the moon than it would on an airless body. Rhea (upper left) appears rough because its icy surface is heavily cratered. A close inspection of Mimas (center bottom), though difficult to see at this scale, shows surface irregularities due to its own violent history. Image Credit: NASA/JPL-Caltech/Space Science Institute

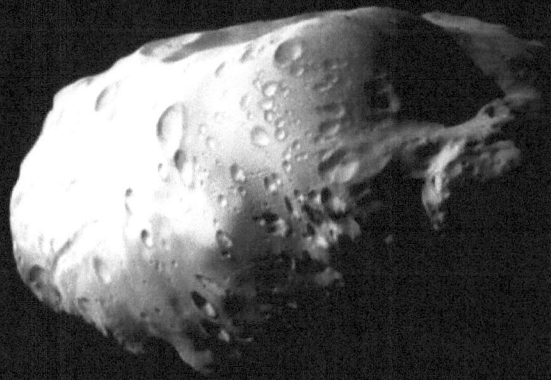

PROMETHEUS
UP CLOSE

Cassini spied details on the pockmarked surface of Saturn's moon Prometheus (53 miles, or 86 kilometers, across) during a flyby on December 6, 2015. This is one of Cassini's highest resolution views of Prometheus, and the view looks toward the side of Prometheus that faces away from Saturn. North on Prometheus is up. Prometheus orbits Saturn just interior to the narrow F ring, which is seen here at top. Image Credit: NASA/JPL-Caltech/Space Science Institute

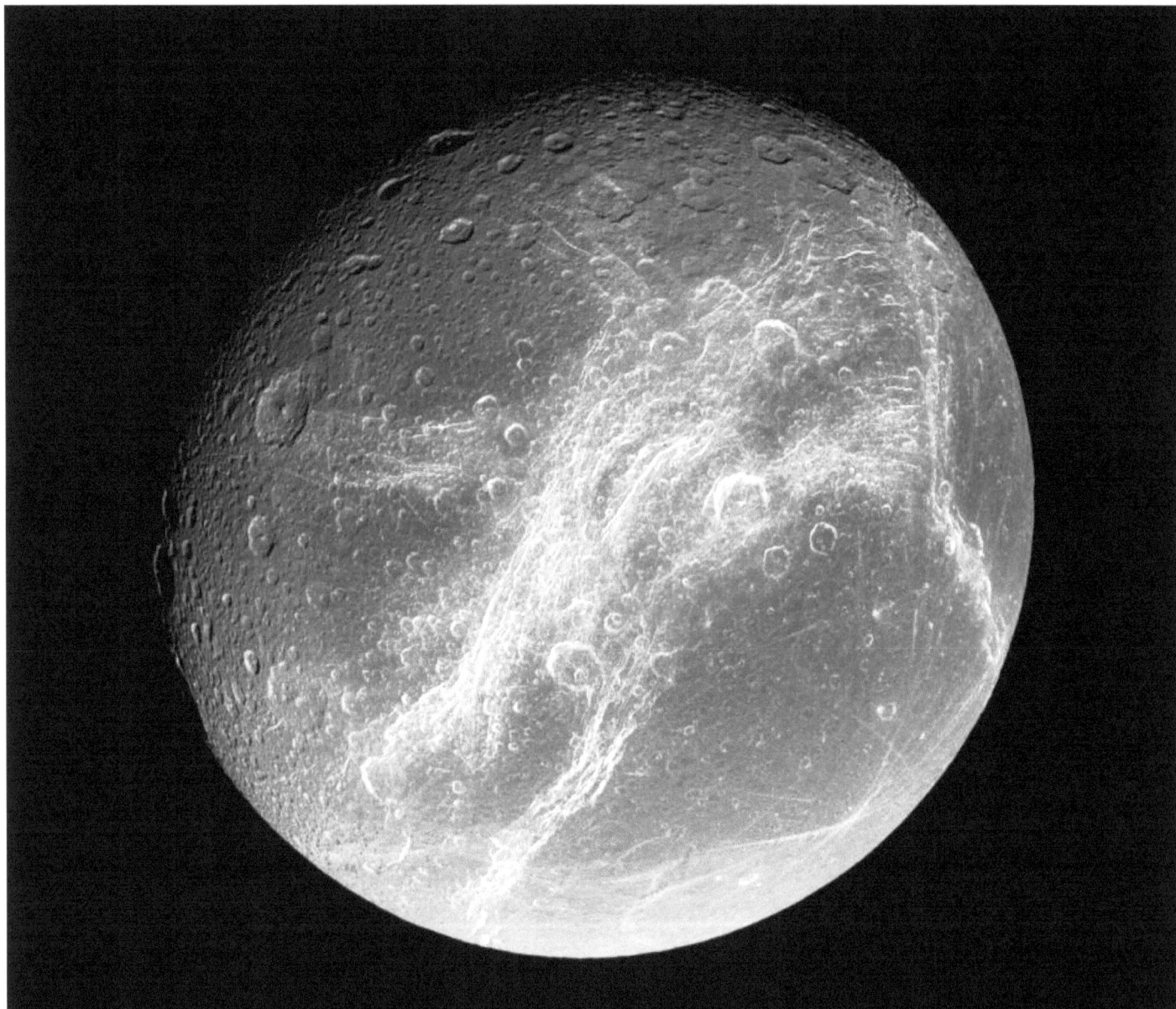

INCREDIBLE CLIFFS

This splendid view showcases Dione's tortured complex of bright cliffs. At lower right is the feature called Cassandra, exhibiting linear rays extending in multiple directions. The trailing hemisphere of Dione (700 miles, or 1,126 kilometers, across) is seen here. North is up. Cassini captured this image in polarized green light on July 24, 2006, at a distance of approximately 163,000 miles (263,000 kilometers) from Dione. Image Credit: NASA/JPL-Caltech/Space Science Institute

EPIMETHEUS REVEALED

Cassini's close flyby of Epimetheus in December 2007 returned detailed images of the moon's south polar region. The view shows what might be the remains of a large impact crater covering most of this face, and which could be responsible for the somewhat flattened shape of the southern part of Epimetheus (72 miles, or 116 kilometers, across) seen previously at much lower resolution. The image also shows two terrain types: darker, smoother areas and brighter, fractured terrain. One interpretation of this image is that the darker material evidently moves down slopes and probably has a lower ice content than the brighter material, which appears more like "bedrock." Nonetheless, materials in both terrains are likely to be rich in water ice. Image Credit: NASA/JPL-Caltech/Space Science Institute

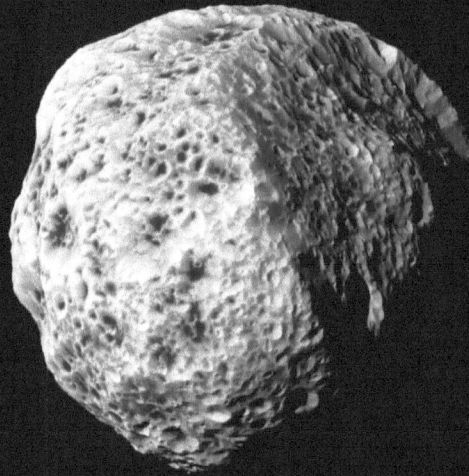

SPONGY HYPERION

The sponge-like surface of Saturn's moon, Hyperion, is highlighted in this Cassini portrait, captured during the spacecraft's flyby on September 16, 2011. Hyperion (168 miles, or 270 kilometers, across) has an irregular shape, and it tumbles through its orbit (that is, it doesn't spin at a constant rate or in a constant orientation). Images such as this one allow a better inventory of the surface features, the satellite's shape, and changes in Hyperion's spin. Image Credit: NASA/JPL-Caltech/Space Science Institute

ADRIFT AT SATURN

Tethys floats before the massive, golden-hued globe of Saturn in this natural-color view, captured by Cassini on December 3, 2005, approximately 1.6 million miles (2.5 million kilometers) from Saturn. The thin, dark line of the rings curve around the horizon at top. Visible on Tethys (665 miles, or 1,071 kilometers, across) are the craters Odysseus (top) and Melanthius (bottom). Cassini took images using red, green, and blue spectral filters, which were combined to create this color view. Tethys is apparently darker than Saturn at these wavelengths. The edge of the planet appears fuzzy, which may indicate that we are seeing haze layers that are separated from the main cloud deck. Image Credit: NASA/JPL-Caltech/Space Science Institute

DAZZLING COLOR

Cool and icy Dione floats in front of giant Saturn bedecked in a dazzling array of colors. The surface of Dione, which exhibits contrasting bright and dark areas when viewed up close, appears pale in this image. But Saturn's multi-hued cloud bands boldly steal the show. Discrete clouds and eddies in Saturn's northern hemisphere can be seen within the faint shadows of the rings on the planet. Dione is 695 miles (1,118 kilometers) across, and Cassini took images using red, green, and blue filters on September 22, 2005, at approximately 499,000 miles (803,000 kilometers) from Dione to create this natural-color view. Image Credit: NASA/JPL-Caltech/Space Science Institute

HIGH-RES HELENE

The Cassini spacecraft snapped this image of Saturn's moon Helene while completing the mission's second-closest encounter of the moon on June 18, 2011. This flyby provided some of the highest-resolution images of the moon. Lit terrain seen here is on the leading hemisphere of Helene (21 miles, or 33 kilometers, across), and north is up. Cassini captured this image approximately 4,000 miles (7,000 kilometers) away. Image Credit: NASA/JPL-Caltech/Space Science Institute

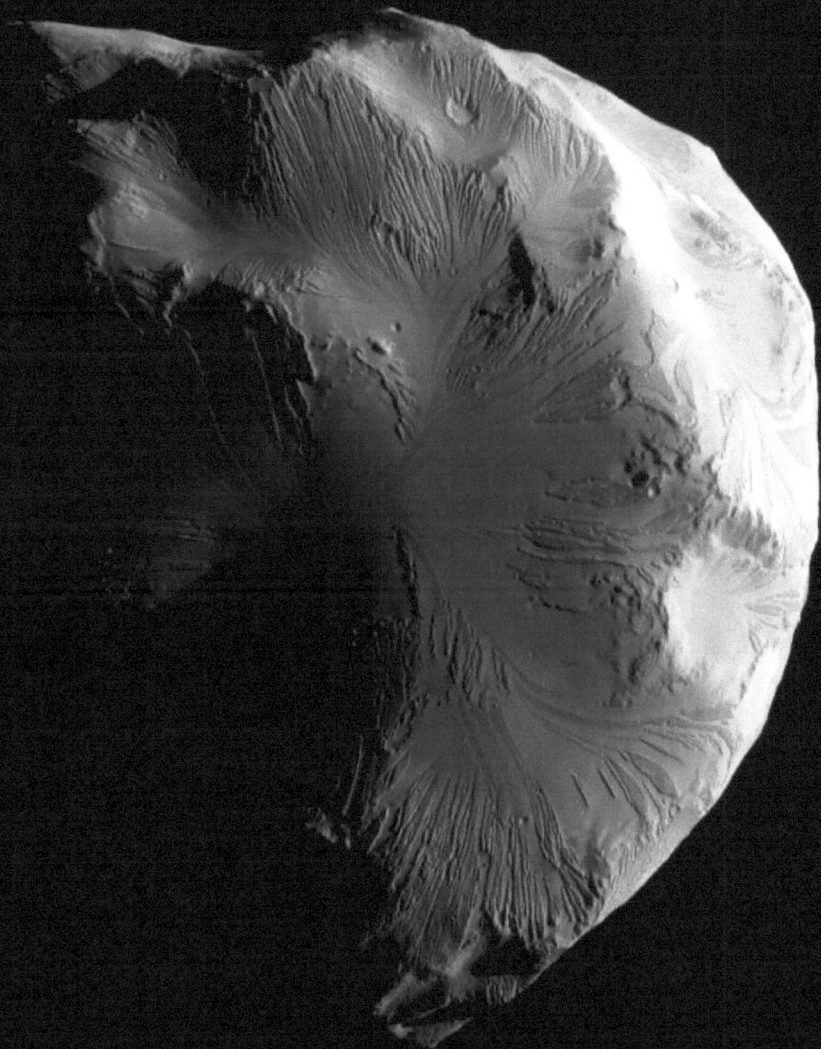

ODD WORLD (HYPERION)

This stunning false-color view of Saturn's moon, Hyperion, reveals crisp details across the strange, tumbling moon's surface. Differences in color could represent differences in the surface composition. Cassini captured this image during a close flyby on September 26, 2005, from approximately 38,500 miles (62,000 kilometers) away. Cassini scientists think that Hyperion's unusual appearance can be attributed to the fact that it has an unusually low density for such a large object, giving it weak surface gravity and high porosity. These characteristics help preserve the original shapes of Hyperion's craters by limiting the amount of impact ejecta coating the moon's surface. Image Credit: NASA/JPL-Caltech/Space Science Institute

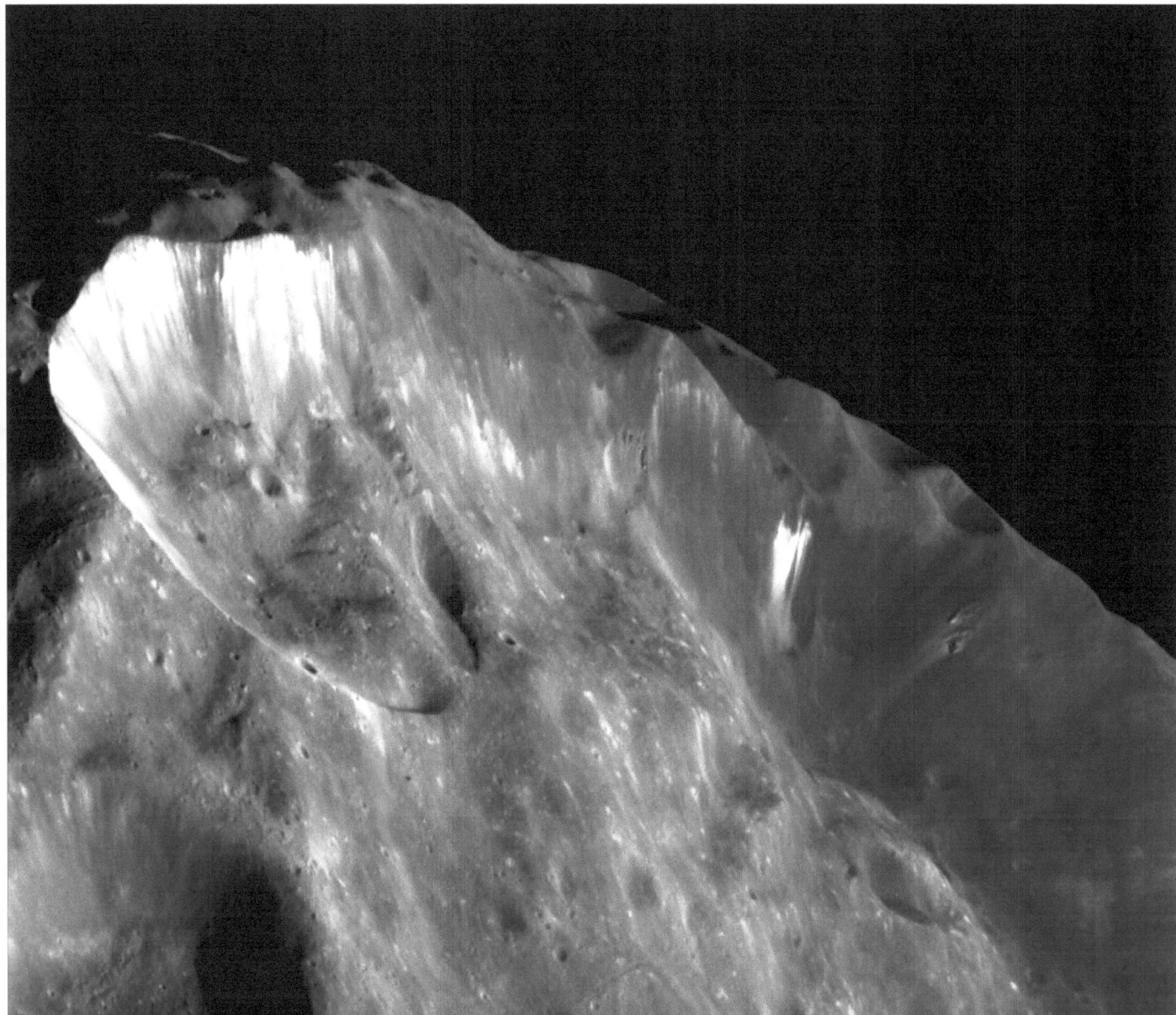

A SKYLINE VIEW (PHOEBE)

This image of Phoebe shows bright wispy streaks thought to be ice. The crater on the left, with most of the bright streaks, is about 28 miles (45 kilometers) in diameter, measured from the side nearest the viewer to the side farthest from the viewer. Cassini captured this image at about 7,400 miles (11,900 kilometers) from Phoebe. No enhancement was performed on this image. Image Credit: NASA/JPL-Caltech/Space Science Institute

THE BIG ONE (MIMAS WITH HERSCHEL)

Mimas' gigantic crater, Herschel, lies near the moon's limb in this Cassini view. A big enough impact could potentially break up a moon. Luckily for Mimas, whatever created Herschel was not quite big enough to do so. When large impacts happen, they deliver tremendous amounts of energy. Even impacts that are not catastrophic can leave enormous, near-permanent scars on bodies like Mimas (246 miles, or 396 kilometers, across). Cassini took this image on November 19, 2016. Image Credit: NASA/JPL-Caltech/Space Science Institute

RED ARCS
ON TETHYS

Unusual arc-shaped, reddish streaks cut across the surface of Saturn's ice-rich moon, Tethys, in this enhanced-color mosaic. The red streaks are narrow, curved lines on the moon's surface, only a few miles (or kilometers) wide but several hundred miles (or kilometers) long. They are among the most unusual color features on Saturn's moons to be revealed by Cassini's cameras. The origin of the features and their reddish color is currently a mystery to Cassini scientists. Possibilities being studied include ideas that the reddish material is exposed ice with chemical impurities or the result of outgassing from inside Tethys. Except for a few small craters on Dione, reddish tinted features are rare on other moons of Saturn. Image Credit: NASA/JPL-Caltech/ Space Science Institute

INTERACTIVE MODELS OF SATURNIAN MOONS

These interactive visualizations of a few of Saturn's moons, Tethys, Dione, Iapetus, and Mimas, were created from images captured by the Cassini spacecraft. For each moon model, multiple images were then stitched together to create a mosaic of the moon's surface, and that mosaic was then projected onto the model.

Select an image below to interact with the model.

INTERACTIVE 5.1 **Tethys**

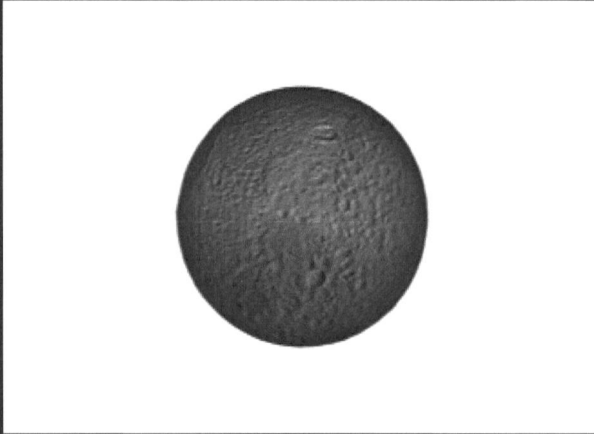

INTERACTIVE 5.3 **Iapetus**

INTERACTIVE 5.2 **Dione**

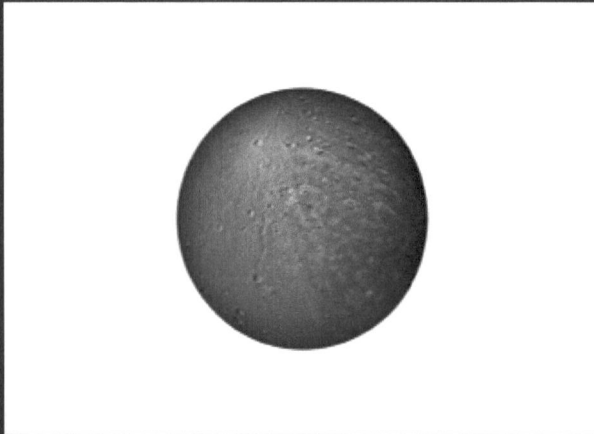

INTERACTIVE 5.4 **Mimas**

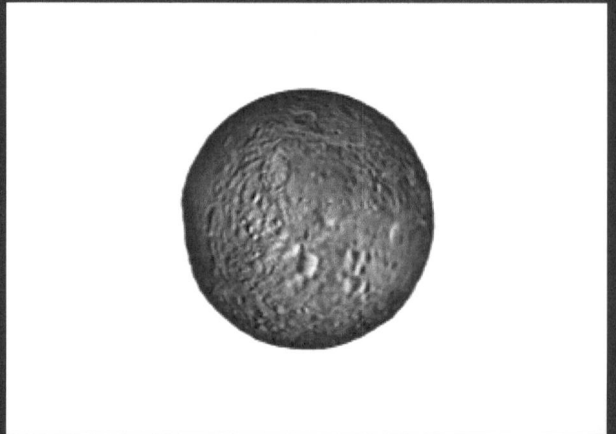

QUINTET

A quintet of Saturn's moons come together in the Cassini spacecraft's field of view for this portrait. Janus (111 miles, or 179 kilometers, across) is on the far left. Pandora (50 miles, or 81 kilometers, across) orbits between the A ring and the thin F ring near the middle of the image. Brightly reflective Enceladus (313 miles, or 504 kilometers, across) appears above the center of the image. Saturn's second largest moon, Rhea (949 miles, or 1,528 kilometers, across), is bisected by the right edge of the image. The smaller moon Mimas (246 miles, or 396 kilometers, across) can be seen beyond Rhea also on the right side of the image. Image Credit: NASA/JPL-Caltech/Space Science Institute

CLOSEST DIONE FLYBY

Flying past Saturn's moon, Dione, Cassini captured this view on December 12, 2011, which includes two smaller moons, Epimetheus and Prometheus, near Saturn's rings. This encounter was the spacecraft's closest pass of Dione's surface, but because this flyby was intended primarily for instruments other than the cameras, it didn't provide Cassini's best images of the moon. Dione (698 miles, or 1,123 kilometers across) is closest to Cassini here and is on the left of the image. Potato-shaped Prometheus (53 miles, or 86 kilometers across) appears above the rings near the center top of the image. Epimetheus (70 miles, or 113 kilometers across) is on the right. Image Credit: NASA/JPL-Caltech/Space Science Institute

MIMAS BLUES

Mimas drifts along in its orbit against the azure backdrop of Saturn's northern latitudes in this true-color view. The long, dark lines on the atmosphere are shadows cast by the planet's rings. Saturn's northern hemisphere was relatively cloud-free at the time the image was taken, and rays of sunlight took a long path through the atmosphere. This resulted in sunlight being scattered at shorter (bluer) wavelengths, thus giving the northernmost latitudes their bluish appearance at visible wavelengths. At the bottom, craters on icy Mimas (247 miles, or 398 kilometers, across) give the moon a dimpled appearance. Image Credit: NASA/JPL-Caltech/Space Science Institute

AFTERWORD
BY EARL MAIZE

CASSINI PROGRAM MANAGER,
NASA'S
JET PROPULSION LABORATORY

OCTOBER 2017

Cassini's mission began on the evening of October 15, 1997 with a fiery departure from the Earth aboard a U.S. Titan IVB/Centaur rocket. Nearly 20 years later, at about 10:45 a.m. Saturn local time on September 15, 2017, Cassini's mission ended as it had begun, this time with a fiery plunge into Saturn's atmosphere.

As a fitting end to 13 incredible years of exploring the Saturn system, the mission concluded with the Grand Finale, a gambit so daring that it could not have been considered during any other phase of the mission. These 22 orbits darted through the gap between the rings and the planet itself, revealing more unseen wonders and a collection of brand-new puzzles.

Christened a "discovery machine" by its scientists, the Cassini mission will be recorded as one of the most productive chapters in humankind's exploration of the planets. From the moment Cassini began science operations until its final moments as it was overcome by Saturn's encroaching atmosphere, Cassini was returning invaluable information about the Saturn system. Time and time again, Cassini shocked and amazed both the science community and the public.

The images contained here are among the most striking from the Cassini mission and represent the ultimate synergy between engineering capability and scientific curiosity. While they are the tip of the iceberg, they are the most overwhelming evidence of the tremendous scientific bounty of the Cassini mission, much of which has been captured in the nearly 4,000 scientific papers from Cassini scientists.

The Cassini mission has ended high over the clouds of Saturn. The Cassini spacecraft is gone; thank you and farewell, faithful explorer. But the Cassini legacy has just begun. The impact that Cassini has on the future of planetary exploration will go on for decades. Long live Cassini.

APPENDIX I:
THE MISSION

Humans have yet to achieve crewed interplanetary travel. Instead, for our most distant missions of exploration, we dispatch un-crewed spacecraft — electronic emmissaries — on reconnaissance missions to the places we cannot venture ourselves. The Cassini-Huygens mission is one of the great milestones in that effort.

It had been 40 years since humans began sending probes into space when Cassini-Huygens launched on October 15, 1997. As a result, the spacecraft was built in an era in which, for the first time, some of the people involved had spent entire careers in the endeavor of spaceflight. This mission was forged from the successes and failures of previous missions that aimed for Earth orbit, lunar landings, early exploration of Mars, and flybys of the outer planets. Spaceflight had begun to mature, and it showed in the design of Cassini, which was equipped with backups of nearly everything — computer, radio antenna, data recorder, thruster system, and main engine, among other systems.

At 22 feet (6.7 meters) high, 13.1 feet (4 meters) wide, and weighing 12,593 pounds (5,712 kg), this masterwork of engineering left Earth in an early-morning launch atop the most powerful rocket available at the time. The rocket roared and crackled into the sky from the marshes of Florida's Cape Canaveral, puncturing and illuminating the clouds before soaring beyond our atmosphere at many thousands of miles per hour. After detaching from the rocket, Cassini raced around the Sun, flying by Venus twice, Earth once, and Jupiter once, each time gaining thousands of miles per hour of velocity.

After Cassini entered Saturn orbit, the spacecraft operations teams and science teams grew more familiar with the spacecraft, as well as with one another. Cassini's route grew increasingly bold, and the scientific observations (including imaging) grew more refined and polished. The mission proved itself to be interplanetary exploration at its best.

Scientific discovery was Cassini's purpose, but the spacecraft itself, along with the brilliant minds that built and operated it, were the means to that ends. This now beloved spacecraft ended its epic quest in, perhaps, the most fitting way it could, by becoming literally part of the planet that it so thoroughly revealed to us — Saturn, the crown jewel of the solar system.

Appendix Title Image:
Preflight Testing

This image shows the newly assembled Cassini spacecraft on October 31, 1996, at the JPL facilities in Pasadena, California, where it was undergoing vibration and thermal testing. The Huygens probe, provided by the European Space Agency (ESA), is the enormous shield-shaped object facing the viewer, and the radio antenna (the huge white dish at top) was provided by the Italian Space Agency (ASI). Image Credit: NASA/JPL-Caltech/Space Science Institute

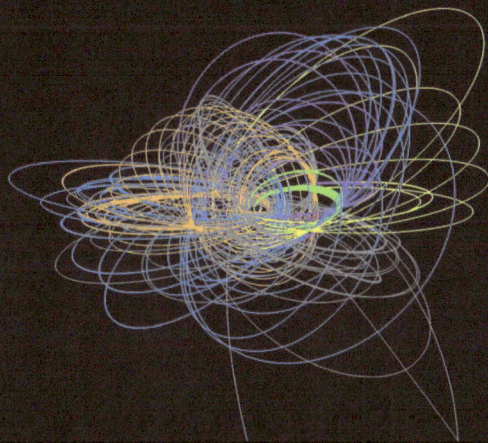

BALL OF YARN

This computer-generated animation shows all of Cassini's orbits of Saturn, which began in 2004 and ended in 2017 — affectionately called the "ball of yarn" by the mission. Image credit: NASA/JPL-Caltech

Select the image above to view a movie of it rotating.

APPENDIX II:
SCIENTIFIC INSTRUMENTS

The Imaging Science Subsystem (ISS)

The Imaging Science Subsystem consisted of a wide-angle and a narrow-angle digital camera. The cameras were sensitive to visible wavelengths of light and to some infrared and ultraviolet wavelengths. Each camera had several filters mounted on wheels to select the wavelengths to be sampled in each image. The Imaging Science Subsystem served as Cassini's main set of eyes for viewing the Saturn system. It studied features as small as boulders and as large as Saturn's rings. The instrument also assisted in navigation and captured many of the beautiful images for which Cassini is so well known.

To find out more, visit https://saturn.jpl.nasa.gov/imaging-science-subsystem/.

Visual and Infrared Mapping Spectrometer (VIMS)

The Visible and Infrared Mapping Spectrometer (VIMS) collected both light that is visible to humans and infrared light of slightly longer wavelengths. It separated the light into its various wavelengths, so scientists can learn about the composition of materials from which the light is reflected or emitted. Scientists used VIMS to determine the content and temperatures of atmospheres, rings, and surfaces in the Saturn system. The instrument analyzed light, but scientists can also create images from its data, similar to a visible-light camera.

To find out more, visit https://saturn.jpl.nasa.gov/visible-and-infrared-mapping-spectrometer/.

Radar Mapper (RADAR)

Invisible to human eyes, radio waves can penetrate thick and murky atmospheres, and they reflect off hard surfaces. Cassini's radar instrument sent radio waves at various targets, and by recording slight differences in the signal's arrival time and wavelength back at the spacecraft, the instrument created pictures of the landscapes upon which it reflected. Cassini's radar instrument was built primarily for studying Saturn's moon, Titan, which has a thick atmosphere that hides its surface. The instrument bounced radio waves off Titan and captured the reflected waves to see the surface that lay below Titan's haze, such as lakes, rivers, mountains, and dunes. The instrument could detect how smooth or rough a surface is, making it useful for studying Titan's methane seas. Scientists also used it to study Saturn, its rings, and Saturn's other moons.

To find out more, visit https://saturn.jpl.nasa.gov/radio-detection-and-ranging/.

Huygens Instruments

The European Space Agency's Huygens Probe was a unique, advanced spacecraft and a crucial part of the overall Cassini mission to explore Saturn. The probe was about 9 feet (2.7 meters) wide and weighed roughly 700 pounds (318 kilograms). It was built like a shellfish: a hard shell protected its delicate interior from high temperatures during the a two-hour-and-27-minute descent through the atmosphere of Saturn's giant moon, Titan. The images in this e-book from the Huygens probe were taken by the probe's Descent Imager/Spectral Radiometer, one of two NASA instruments on the probe. The probe landed on Titan on January 14, 2005, becoming the first human-made spacecraft to land in the outer solar system.

To find out more, visit https://saturn.jpl.nasa.gov/mission/spacecraft/huygens-probe/.

Appendix Title Image:
Making Adjustments

On September 13, 1997, one month and two days before Cassini launched, engineers work on the spacecraft at NASA's Kennedy Space Center. Image Credit: NASA